DEDICATED

To My

Late Grandmother

&

Venerable Parents and Teachers

Preface

The endless thanks go to **Lord Ganesha and Maa Saraswati ji as Unseen Power** for all the blessings he has showered on to me, which has enabled me to write this work in my target.

Genetics is a branch of biological science, which has wide scope of their implementation in both animal and plant kingdom. In relation to genetics, there are many competitive books available in market for different examinations conducted by Indian Council of Agricultural Research, Agriculture Scientist Recruitment Board and others State or Central Universities for admission and services purposes. It is astonishing that there is hardly any book available in the market on genetics which contains basic to molecular aspects to meet the needs of the student particularly who preparing JRF, undergraduate or post graduate examination. Thus the students require to hard work for collection of needful study materials.

In keeping of aforesaid situations, this book entitled **Objective Genetics, Basic to Molecular** has been designed on persistent demand of the Indian students. This book has been consisted fifteen chapters based the courses offered by State/Central Agriculture Universities and others institutions related life sciences. Each chapter of this book has been covered terminology and objectives type questions. This book is also cover important point views asked by experts to candidate in different Government or Private Organizations Services.

I express my every gratitude and most heartily loyalty to my respected parents Shri Rambhajan Upadhyay and Smt. Kanta Devi who set the underpinning and were always with me during my ups and downs. I also express my special regards to my grandfather Shri Raghuveer Upadhyay and all my family members and relatives for their blessings and constant encouragement during this target.

I am also thankful to Dr. Lokendra Singh, Associate Professor and Dr. S. K. Singh Assistant Professor, Department of Genetics and Plant Breeding of C. S. Azad University of Agriculture and Technology, Kanpur and Dr. Kamlesh Kumar, Ex-Professor and Dr. Shiva Nath, Assistant Professor, Department of Genetics and Plant Breeding of N. D. University of agriculture and Technology, Kumarganj Faizabad, Uttar Pradesh (India), Dr. Omkar Singh (Botany), Dr. P. K. Singh (Agronomy), Dr. Rajani Rani (Bio-Chemistry), R. K. (P.G.) College, Shamli and Shri Puspendra Kumar Dalaal, Research Scholar (Agronomy), C.S.A.U.A.&T., Kanpur, Uttar Pradesh (India) for encouraged me in my creative work.

All reader, researchers and students are requested to put their valuable suggestions for further improving the book in future.

Jaydev Kumar

Contents

Chapter **1**

Cell and Cell Division

1. Who discovered the cell in 1665?
   A. Robert Brown B. Gill
   C. Robert Hook D. None
2. The theory of cell was proposed by :
   A. Blackslee B. Correns
   C. Schwann and Schleiden D. Rudolf Virchow
3. When did, Schwann and Schleiden propose cell theory?
   A. 1839 B. 1876
   C. 1865 D. 1665
4. Who suggested cell as the basic unit of life in 1839?
   A. Schwann and Schleiden B. Rudolf Virchow
   C. Robert Remak D. Henri Dutrochet
5. I[st] compound microscope was made by :
   A. Zacharias Jansen B. Anton van Leeuwenhoek
   C. Both A & B D. None
6. Green plants are also known as :
   A. Viridiphyta B. Viridiplantae
   C. Chlorobionta D. All of these
7. A disomic cell is :
   A. Somatic cell B. Diploid cell
   C. Normal cell D. All of these
8. Middle lamella of cell wall consists of :
   A. Pectin B. Cellulose
   C. Carbohydrates D. None
9. Which of the following materials are the compositions of pectin?
   A. Ca + Mg B. Mg + Cu
   C. Both A & B D. None

10. In plant, cell wall found between plasma membrane and middle lamella:
    A. Primary cell wall
    B. Secondary cell wall
    C. Both A & B
    D. None
11. Which of the following is not found in animal cell?
    A. Centrioles
    B. Vacuoles
    C. Mitochondria
    D. All of these
12. Constitution of algae cell :
    A. Hemicelluloses, pectins and protein
    B. Pectins, protein and cellulose
    C. Cellulose, hemicellulose and pectin
    D. Cellulose, galactans and mannans
13. Which of the following single celled individual do come in eukaryotes?
    A. Protista
    B. Monera
    C. Fungi
    D. Archaea
14. Which of the following is a major compound of primary cell wall?
    A. Hemi-cellulose
    B. Cellulose
    C. Pectin
    D. None
15. Cell wall is absent in :
    A. Plant
    B. Animals
    C. Both A & B
    D. None
16. Which of the following is the structural unit of cell wall?
    A. Cellulose and microfibril
    B. Protein
    C. Lipids
    D. None
17. Secondary cell wall mainly consists of :
    A. Cellulose
    B. Hemicelluloses
    C. Cutin
    D. All of these
18. Plasmalemma consists of :
    A. Protein and lipids
    B. Microfibrils
    C. Both A & B
    D. None
19. In plasmalemma, major proportion of lipids consists of :
    A. Phospholipids
    B. Protein
    C. Carbohydrates
    D. All of these
20. In 1935, the lipid-protein bi-layer model as the unit of membrane structure was proposed by :
    A. Davson and Danieli
    B. Holiker
    C. Johannson
    D. Dave
21. Lecithin is a :
    A. Lipid
    B. Phospholipid
    C. Carbohydrates
    D. None

22. Insect cell is made up of :
A. Cellulose B. Chitin
C. Protein D. Hemicelluloses

23. Materials of cytoplasm except other than structure is:
A. Mitochondria B. E. R.
C. Hyaloplasm D. All of these

24. Centriole is found in :
A. Animal cell B. Plant cell
C. Both A & B D. None

25. Prokaryotic organization body is constricted of :
A. Unicellular cell B. Multi-cellular cell
C. Both A & B D. None

26. Endoplasmic reticulum is absent in :
A. Eukaryotic cell B. Prokaryotic cell
C. Both A & B D. None

27. Protein synthesis occurs in :
A. Smooth type E. R. B. Rough type E. R.
C. Golgi bodies D. Mitochondria

28. Protein synthesis takes place in :
A. Ribosome B. Golgi bodies
C. Nucleus D. None

29. In plant, which of the following process requires sulphur :
A. ATP synthesis B. Protein synthesis
C. Glucose synthesis D. DNA replication

30. Which one of the following techniques is used to develop a disease and insect free plant?
A. Embryo culture B. Suspension culture
C. Meristem culture D. Anther culture

31. E. R. is absent in :
A. RBCs B. Egg cell
C. Undifferentiated cell D. All of these

32. Major function of E. R. is :
A. Sorting of protein B. Modification of protein
C. Both A & B D. None

33. Eukaryotic cell contains :
A. 80S Ribosome B. 70S Ribosome
C. 100S Ribosome D. None

34. Protein synthesis does not occur in :
    A. Smooth type E. R.
    B. Rough type E. R.
    C. Both A & B
    D. None
35. Nucleus is absent in :
    A. Prokaryotic cell
    B. Eukaryotic cell
    C. Both A & B
    D. None
36. Ribosome is composed of :
    A. RNA
    B. Protein
    C. Both A & B
    D. None
37. Ribosomes are present on :
    A. Outer surface of smooth type E. R.
    B. Golgi bodies
    C. Outer surface of rough type E. R.
    D. None
38. Site of protein synthesis is :
    A. Mitochondria
    B. Centrioles
    C. Ribosome
    D. All of these
39. In animal and plant, ribosome are composed of :
    A. 55-60% protein
    B. 40-45% RNA
    C. Both A & B
    D. None
40. Ribosomes are synthesized in :
    A. Nucleolus
    B. Cytoplasm
    C. Both A & B
    D. None
41. Ribosome is assembled in :
    A. Nucleolus
    B. Cytoplasm
    C. Both A & B
    D. None
42. Ribosome are synthesized and assembled in nucleolus within :
    A. Cytoplasm
    B. Hyaloplasm
    C. Nucleus
    D. None
43. Polysome consists of :
    A. 5-8 ribosomes during protein synthesis
    B. Only two ribosome mRNA formation
    C. Both A & B
    D. None
44. In plant or animal, size of macromolecules is expressed by :
    A. 'N' unit
    B. 'S' unit or Svedberg
    C. 'R' unit
    D. None
45. Golgi complex was 1st of all described by :
    A. Golgi
    B. Camillo Golgi
    C. Muller
    D. Morgan

46. Golgi bodies was 1st described in :

A. Nerve cells of cat
B. Nerve cells of owl
C. Both A & B
D. None

47. In which year, Golgi apparatus was discovered by C. Golgi ?

A. 1882
B. 1822
C. 1977
D. 1922

48. Golgi complex arises, in which of the following :

A. Rough E. R.
B. Smooth E. R.
C. Mitochondria
D. All of these

49. Major function of golgi complex is :

A. Packaging of food materials
B. Oxidation of carbohydrate
C. Provide the ATP for cell
D. None

50. Golgi complex is responsible for the production of :

A. Chromosome
B. Lysosome
C. Chloroplast
D. None

51. Lysosome was discovered by :

A. Duve
B. Jones
C. Camillo Golgi
D. All of these

52. In which year, Lysosome was discovered by Duve?

A. 1951
B. 1949
C. 1761
D. 1955

53. Which of the following types of lysosome is produced by golgi complex?

A. Primary lysosome
B. Secondary lysosome
C. Tertiary lysosome
D. None

54. Which of the following types of lysosome contains only hydrolytic enzyme?

A. Secondary lysosome
B. Primary lysosome
C. Tertiary lysosome
D. None

55. Secondary lysosome contains :

A. Food materials
B. Hydrolytic enzyme
C. Both A & B
D. None

56. Lysosome is a term which has meaning one of the following:

A. Some = body
B. Lyse = destroy
C. Both A & B
D. None

57. Which of the following part of a cell is known as suicidal bags?

A. Lysosome
B. Stroma
C. Grana
D. None

58. Peroxisomes are found in :
   A. Animal B. Plant
   C. Both A & B D. None
59. Spherosome is found in :
   A. Plant B. Animal
   C. Both A & B D. None
60. Major function of spherosome is :
   A. Storage of carbohydrates B. Storage of nucleic acid
   C. Storage of lipids D. All of these
61. In plant cell, peroxisomes was reported by :
   A. Talbert *et al.,* 1968 B. Hill, 1989
   C. Z. Muller D. None
62. The term mitochondria was coined by :
   A. Benda, 1897 B. Venda, 1896
   C. Jenssen, 1997 D. Kumar, 2000
63. Which of the following part of nucleus is responsible for generation of mitochondria?
   A. Nucleolus B. Nuclear envelope
   C. Chromatin D. None of these
64. Mitochondria was discovered by :
   A. Holiker B. Benda
   C. Holiker and Benda D. None
65. Mitochondria consists of :
   A. Protein B. Lipids and RNA
   C. DNA D. All of these
66. Which of the following is a part of mitochondria?
   A. Membrane B. Cristae
   C. Matrix D. All of these
67. Mitochondria has :
   A. Circular DNA B. Linear DNA
   C. Both D. None
68. Bactera "*Borrelia burgdorferi*" has :
   A. Circular DNA B. Linear DNA
   C. Both A & B D. None
69. An extra chromosomal DNA in a bacterial cell is :
   A. Plasmid B. DNA
   C. Chromosome D. None

70. The bacterial cell wall differs from that of all other organisms by the presence of :

A. Hemicelluloses
B. Peptidoglycan
C. Cellulose
D. All of these

71. Both gram-positive and gram-negative bacteria have a common membrane :

A. S-layer
B. N-layer
C. Both A & B
D. C-layer

72. In gram-positive bacteria, S-layer is attached to :

A. Outer membrane
B. Peptidoglycan layer
C. Both A & B
D. None

73. The S-layer is attached directly to the outer membrane in :

A. Gram positive bacteria
B. Gram negative bacteria
C. Both A & B
D. None

74. The term "periplasm" was coined by :

A. Mitchell (1961)
B. Mc Clintock (1983)
C. Mannan *et al.* (1905)
D. None

75. In mitochondria, space between outer and inner membranes is :

A. 50-57^{0}A
B. 40-47^0 A
C. 30-37^{0}A
D. None

76. Which of the following is known as power house of cell?

A. Golgi bodies
B. RBCs
C. Mitochondria
D. None

77. Which of the following, part of cell has self replication capacity?

A. Mitochondria
B. Chloroplast
C. DNA
D. All of these

78. Mitochondria is originated from :

A. Nucleus
B. Nuclear envelope
C. Nucleolus
D. None

79. The site of cell respiration is :

A. Mitochondria
B. Chloroplast
C. E. R
D. All of these

80. Mitochondria contains :

A. Oxidative enzymes
B. ATP
C. Both A & B
D. None

81. Mitochondrial genes responsible for :

A. Nuclear inheritance
B. Plasma inheritance or cyto-plasmic inheritance
C. Both A & B
D. None

82. The energy produced in mitochondria is stored in the form of :

A. ATP B. Carbohydrates

C. Lipids D. Protein

83. Splitting of water is associated with which of the following process of photosynthesis :

A. Photosystem II B. Photosystem I

C. Chlorophyll b D. Xanthophylls

84. Kreb's cycle occurs in :

A. Matrix of mitochondria B. E. R.

C. Chloroplast D. None

85. Who discovered Kreb's cycle?

A. Emmerson B. Hill

C. Hans Adolf Kreb D. None

86. Total energy produced in Kreb's cycle is equal to :

A. 2 ATP B. 24 ATP

C. 20 ATP D. 18 ATP

87. TCA or Krebs cycle is a enzymatic process and their enzymes are mainly found in :

A. Matrix of mitochondria B. E. R.

C. Golgi bodies D. Chloroplast

88. How many steps involve in TCA cycle?

A. Ten B. Five

C. Fifteen D. All of these

89. Kreb's cycle is also known as :

A. Citric cycle B. Tricarboxylic acid cycle (TCA)

C. Both A & B D. None

90. Which of the following is the main source for electrons transport chain in mitochondria?

A. NADH B. ATP

C. ADP D. None

91. Plastids are absents in :

A. Bacteria B. Fungi

C. Animals D. All of these

92. Plastids are present in :

A. Plant B. Animal

C. Both A & B D. None

93. Which of the following part (s) have capacity of self replication ?

A. Plastid B. Mitochondria

C. Both A & B D. None

94. Which of the following plastid store the starch, fat and protein ?

A. Leucoplast B. Chromoplast

C. Chloroplast D. All of these

95. The colorless plastid is called as :

A. Leucoplast B. Chromoplast

C. Chloroplast D. All of these

96. Chloroplast has :

A. Circular DNA B. Linear DNA

C. Both A & B D. None

97. Chromoplast is a type of plastid which contains different types of color except :

A. Green color B. Blue color

C. Red color D. All of these

98. Site of photosynthesis in green plant is :

A. Mitochondria B. Chromosome

C. Chloroplast D. All of these

99. Grana and stroma are the parts of :

A. Nucleus B. Cell

C. Chloroplast D. All of these

100. Which of the following, functions is controlled by leucoplast ?

A. Storage of protein and fat B. Storage of starch

C. Plastid contain green color D. Both A & B

101. Which of the following compounds is a major portion of chloroplast?

A. Protein B. Chlorophyll

C. Lipids D. All of these

102. Which of the following colors or pigments is useful in photosynthesis?

A. Chlorophyll B. Carotenoids

C. Xanthophylls D. All of these

103. Which of the following organs of plant consists photosystem II?

A. Grana B. Stroma

C. Entire chloroplast D. Chloroplast membrane

104. Chlorophyll consists of :

A. A head of phytol and tail of pyrrole ring

B. A head of linked carbons and tail of four pyrrole ring

C. A head of four pyrrole ring and tail of linked nitrogens

D. A head of four pyrrole rings and an alchoholic phytol tail

105. Which of the following pigment is water soluble?

A. Chlorophyll
B. Xanthophylls
C. Phycobilin
D. Cerotenoids

106. Circular DNA (deoxyribonucleic acid) is observed in :

A. Chloroplast
B. Mitochondria
C. Both A & B
D. None

107. Group of stroma and grana lamelle is known as :

A. Quantasomes
B. Quanta
C. Inner membrane
D. None

108. Which of the following reactions occurs in grana?

A. Dark reaction
B. Light reaction
C. Both A & B
D. None

109. The universal energy currency of life is :

A. ATP
B. ADP
C. NADPH
D. All of these

110. Which of the following sentence is correct in relation to Redox (reduction-oxidation)?

A. Chemical reactions in which electrons are transferred from a donor molecule to an acceptor molecule
B. Electrons are transferred from acceptor molecules to donor molecules
C. Sources of energy are available to living organisms
D. Both A & C

111. Who coined the term endoplasmic reticulum in 1948?

A. Mc Clintock
B. Muller
C. Porter
D. Kumar S

112. Which of the following terms was given by Mc Clintock in 1934?

A. Mitochondria
B. Nucleolar organizer region
C. Nucleus
D. None

113. Which of the following, scientist has the credit to discover of the nucleus?

A. Jones
B. Robert Brown
C. Benda
D. Sprегue and Tatum

114. Nucleus was discovered in :

A. 1833
B. 1835
C. 1856
D. 1890

115. Which of the following organelle of cell is also known as "controlling centre" of the cell?

A. Mitochondria
B. Golgi bodies
C. Nucleus
D. All of these

116. Nucleus contains :

A. Chromosomes B. Cytoplasm

C. Gene D. Both A & C

117. Major functions of nuclear envelope is :

A. Protect the chromosomes from cytoplasmic effect

B. Permits the transport of e^-

C. Exchange of materials between nucleus and cytoplasm

D. All of these

118. Major function of nucleolus is :

A. To help in formation of ribosome B. Synthesis of RNA

C. Both A & B D. None

119. Chromatin is composed of :

A. Protein B. RNA

C. DNA D. All of these

120. Nucleolus is a part of nucleus, which originates from :

A. NOR or secondary construction B. Primary construction of centromere

C. Telomere D. None

121. The inner portion of the cell without nucleus is known as :

A. Matrix B. Cytoplasm

C. Hyaloplasm D. None

122. Basic unit of any living organism is :

A. Cell B. DNA

C. RNA D. All of these

123. The term chloroplast was proposed in 1885 by :

A. Schimper B. Kotayam

C. Blackslee D. None

124. DNA and RNA are found in :

A. E.R. B. Nucleus

C. Golgi bodies D. All of these

125. Which of the following, term was coined by Strugger in 1950?

A. Proplastid B. Plastid

C. Chloroplast D. All of these

126. In a cell, food materials packaging occurs in :

A. Cytoplasm B. Hyaloplasm

C. Golgi bodies D. None

127. Which of the following, term was coined by Strasburger (1882) ?

A. Cytoplasm
B. Proplastid
C. Mitochondria
D. None

128. The process of production of new cells from the pre-existing cells requires a specific time for its continuity is known as :

A. Cell cycle
B. Cell division
C. Both A & B
D. None

129. A process, in which a parental cell gives rise to new daughter cells is called :

A. Cell division
B. Cell cycle
C. Both A & B
D. None

130. The cell that undergoes cell division is known as :

A. Parent cell
B. Pre-existing cell
C. Daughter cell
D. Both A & B

131. Division of chromosomes and cytoplasm into two daughter cells is called :

A. Cell division
B. Cell cycle
C. Cell synthesis cycle
D. All of these

132. What are the outcomes of cell division ?

A. Production of new cells from a single cell
B. Keep the shape and size of cell within a limited range
C. Formation of new tissues for special work
D. All of these

133. Karyokinesis is related to :

A. Division of cytoplasm into daughter cell from parental cell
B. Division of chromosomes into daughter cell from pre-existing cell
C. Both A & B
D. None

134. Cyto-skeletal elements play crucial roles in :

A. Cell division,
B. Protection
C. Shape and polarity determination in various prokaryotes
D. All of these

135. Karyokinesis refers to :

A. Division of cytoplasm
B. Division of chromosomes
C. Division of cell
D. Both A & B

136. Bacterial cell contains :

A. Circular chromosomes
B. Circular DNA
C. Linear chromosomes
D. Both A & B

137. In bacterial cell, chromosomes are attached with :

A. Plasma lemma B. Secondary cell wall

C. Both A & B D. None

138. In bacteria, replication of chromosomes occurs :

A. Unidirectional B. Bidirectional

C. Only clockwise D. All of these

139. Which of the following term was proposed by Flemming?

A. Meiosis B. Mitosis

C. Chloroplast D. Both A & B

140. Which of the following division produces identical daughter cells in relation to parent cell?

A. Mitotic cell division B. Meiosis cell division

C. Both A & B D. None

141. Which of the following sentence is correct for mitosis?

A. Identical new cells in respect of parent cell

B. Produces two daughter cells from single cell

C. Mitosis occur in somatic cell only

D. All of these

142. First of all, mitotic cell division was described by :

A. Flemming, 1879 B. Albertmann, 1893

C. Maheshwari 1882 D. None

143. Which cell division is responsible for the division of somatic cells?

A. Meiosis B. Endo-mitosis

C. Mitosis D. All of these

144. Endo-mitosis refers to :

A. Division of chromosomes without cell division

B. Division of chromosomes into two daughter cell

C. Both A & B

D. It is a type of meiosis

145. Somatic cell, chromosome number is denoted by :

A. "n" B. 2n or 2x

C. Both A & B D. None

146. Mitosis is helpful in one of the following ways :

A. Regeneration of damaged tissues B. Production of new organs

C. Production of megaspore and microspore D. All of these

147. Cell division occurs in :

A. Germ cell
B. Gametic cell
C. Somatic cell
D. All of these

148. Which of the following phases is perquisite for somatic cell division?

A. Interphase
B. Mitotic phase
C. Meiosis phase
D. Both A & B

149. Which of the following phases is involve an interphase :

A. G_1
B. G_2
C. S
D. All of these

150. Which of the following phase has shortest duration in interphase?

A. G_1
B. S
C. G_2
D. None

151. Which of the following phase comes after prophase in cell division?

A. Anaphase
B. S phase of interphase
C. Telophase
D. None

152. Which of the following phase comes just after S phase of interphase?

A. G_1
B. G_2
C. Both A & B
D. Prophase I

153. Correct order of phases of cell division are :

A. Prophase, Metaphase, Anaphase and Telophase
B. Telophase, Metaphase, Prophase, and Anaphase
C. Metaphase, Prophase, Telophase, and Anaphase
D. Anaphase, Telophase, Metaphase, and Prophase

154. Correct orders of phases involved in cell cycle is :

A. Interphase, prophase, metaphase, anaphase and telophase
B. Anaphase, prophase, Interphase, telophase and metaphase
C. Interphase, telophase, zygotene, cytokinesis and metaphase
D. Telophase, interphase, metaphase, prophase and anaphase

155. Process of RNA and protein synthesis take place at:

A. Prophase of mitosis
B. Metaphase of meiosis
C. Zygotene of prophase
D. G_1 & G_2 phases of interphase

156. The phase, which comes just after prophase in cell cycle is :

A. G_1 phase
B. G_2 phase
C. S phase
D. M phase

157. DNA synthesis occur in :

A. S phase
B. Metaphase
C. G_1 phase
D. G_2 phase

158. A cell pass through G_1, S, G_2 and Mitosis phases, this phenomenon is known as :

A. Cell cycle B. Cell synthesis cycle

C. Cell growth cycle D. All of these

159. Nucleolus appears in which of the following phase of mitosis?

A. Anaphase B. Prophase

C. Both A & B D. Metaphase

160. Nuclear membranes are present in which of the following phase of mitosis?

A. Prophase B. Metaphase

C. Anaphase D. All of these

161. Which of the following events occurs in prophase?

A. Chromosomes are seen like as loose ball of thin wool

B. Relational coiling occurs in chromosomes

C. Nucleolus and nuclear membrane are also present

D. All of these

162. Which of the following enzyme is responsible for super coiling of chromosomes condensing?

A. Topoisomerase I B. Topoisomerase II

C. Ligase D. DNAse

163. How many types of relational coiling are found during cell division?

A. 2 B. 3

C. 4 D. 5

164. Relational coiling are of two types as :

A. Plectonimic coiling B. Paranimic coiling

C. Both A & B D. None

165. Plectonimic coiling is found between :

A. Sister chomatids during metaphase of mitosis

B. Sister chromatids during prophase of mitosis

C. Non-sister chromatids during metaphase of mitosis

D. Non sister chromatids during prophase of mitosis

166. The event occurs in metaphase of mitosis is :

A. Absence of nucleolus B. Disappear of nuclear envelope

C. Absence of relational coiling D. All of these

167. Which of the following features can be seen at metaphase stage?

A. Chromosomes are thicker and short B. Chromosomes are arranged on equatorial plane

C. Spindle fibres are organized D. All of these

168. A process, in which separation of chromosomes into two daughter nuclei within nucleus is called :

A. Closed mitosis B. Open mitosis

C. Meiosis D. All of these

169. A phase, in which two sister chromatids of each chromosome divide and migrate towards opposite poles is known as :

A. Prophase B. Metaphase
C. Telophase D. None

170. A phase, in which two sister chromatids of each chromosome divide and migrate opposite poles is known as :

A. Prophase B. Metaphase
C. Telophase D. Anaphase

171. Spindle fibres originate from :

A. Centromeres B. Telomere
C. NOR D. None

172. Spindle fibres are composed of :

A. Micro tubules B. Carbohydrates
C. Fat D. Lipids

173. Microtubules are made up of :

A. Fat B. Carbohydrate
C. Cellulose D. Proteins

174. Formation of spindle apparatus can be checked by using :

A. Colchicine B. Hydrazine
C. ADP D. Both A & B

175. In which of the following phases of cell division, spindle fibres are not organized?

A. Metaphase B. Prophase
C. Anaphase D. Prophase

176. How many daughter cells are produced in mitotic division?

A. Two B. Three
C. Four D. None

177. Nucleolus originates from :

A. NOR B. Centromere
C. Large chromatid of a homologous chromosome
D. All of these

178. Meiosis first of all observed in :

A. Drosophilla B. Roundworm
C. Mite D. None

179. Who described meiosis in roundworm in year 1883?

A. Morgan B. Ven Beneden
C. Ven mellcol D. Muller

180. Which event, occurs during meiosis?
   A. Production of 'n' gametes
   B. Separation of two or more alleles of each gene
   C. Separation of homologous chromosomes from each other
   D. All of these

181. Meiosis occurs in :
   A. Reproductive cell of plant or animals   B. Tissues developmental phase
   C. Both A & B   D. None

182. The correct sequence of sub-phases of prophase I is :
   A. Leptotene, pachytene, zygotene, diplotene, and diakinesis
   B. pachytene, zygotene, leptotene, diplotene, and diakinesis
   C. pachytene, leptotene zygotene, diplotene, and diakinesis
   D. Leptotene, zygotene, pachytene, diplotene, and diakinesis

183. Which of the following events occur during prophase I of meiosis?
   A. Pairing of homologous chromosomes and crossing over
   B. Condensation of chromosomes and movement of homologous chromosomes away from each other without separation of chromosomes
   C. Formation and terminalization of chiasmata
   D. All of these

184. Which of the following events occur in zygotene of prophase I during meiosis?
   A. Replication of zDNA   B. Pairing of homologous chromosomes
   C. Formation of synaptonemal complex   D. All of these

185. Synaptonemal formation occurs between :
   A. Homologous chromosomes   B. Non homologous chromosomes
   C. Non sister chromatids   D. Both A & C

186. Which, substrate is required for chromosomes condensation during cell division?
   A. Lipids   B. Fat
   C. Proteins   D. All of these

187. Special nuclear protein synthesis occurs at :
   A. Zygotene   B. Leptotene
   C. Pachytene   D. All of these

188. Which of the following events occur during leptotene?
   A. Increment in nucleolus volume   B. Syneptonemal complex organized
   C. Crossing over occurs   D. None of these

189. How many sub-phases of prophase I are in cell division?
   A. Five   B. Six
   C. Four   D. None of these

190. Which of the following event occurs in meiosis?

A. Crossing over and formation of chiasmata

B. Each number of bivalent chromosome has two chromatids

C. Nucleolus present and joined to a chromosome

D. All of these

191. Which of the following event, occurs during diplotene?

A. Decrease of nucleolus in size
B. Crossing over occurs
C. Homologous pairing occurs
D. All of these

192. Mitotic spindle fibres are associated with :

A. Proteins

B. MTOCs (microtubules organizing centre)

C. Both A & B

D. None

193. Protein is involved in the formation of spindle fibres during cell division :

A. α- subunit of the tubulin
B. β-subunit of the tubulin
C. Both A & B
D. None of these

194. The γ-tubulin combines with other proteins to form a circular structure which is known as :

A. The "γ-tubulin ring complex" (γ-TuRC)
B. The "γ-tubulin ring" (γ-TuR)
C. Both A & B
D. None of these

195. Which protein provides stability to microtubules in cell?

A. MAP-4 (microtubules associated protein 4)
B. MAP-1
C. MAP-5
D. MAP-2

196. Who proposed that microtubules use their dynamic properties of growth and shrinkage at their plus ends to probe the three dimensional space of the cell?

A. Marc Kirschner and Tim Mitchison
B. Duve
C. C. Golgi
D. None of these

197. MTOC's are also called :

A. Vacuoles
B. Centrioles
C. Both A & B
D. None of these

198. A little amount of mRNA is synthesized during :

A. G_2
B. S
C. M
D. None of these

199. Various types of recombinations are produced by

A. Meiosis
B. Mitosis
C. Both A & B
D. None

200. Who discovered synaptonemal complex during 1956?
A. Montrose J. Moses
B. Zikler D
C. Karp C
D. None of these

201. "Tripartite structure" is found in syneptonemal complex and can be seen during :
A. Gametogenesis in males
B. Gametogenesis in females
C. Both A & B
D. None of these

202. Tripartite stature of synaptonemal compex is seen during :
A. Pachytene sub stage of prophase I
B. Leptotene sub stage of prophase I
C. Zygotene sub stage of prophase I
D. None of these

203. A full set of sister chromatids is developed in the sub stage of interphase :
A. G_1
B. G_2
C. S
D. M

204. The sister chromatids get separated from each other during :
A. Meiosis II
B. Mitotic cell division
C. Both A & B
D. None of these

205. Crossing over is a process, by which recombinants are produced during :
A. Pachytene
B. Zygotene
C. Prophase II
D. All of these

206. Synaptonemal complexes (SCs) are zipper-like structures, which get assembled between homologous chromosomes during the prophase of :
A. First meiotic division
B. Second meiotic division
C. Endo-mitosis
D. All of these

Terms

- **Anaphase:** During this phase, two chromatids of each chromosome are pulled apart at opposite poles. This phase is known as shortest phase of mitosis.
- **ATP:** Adenosine triphosphate as energy rich component that promotes certain activities in the cell.
- **Cell division cycle or cell cycle:** A process by which a cell produces after duplication genetic material of cell. The four phases of the cell cycle are designated G1, S, G2, and M.
- **Cell Division:** A process, in which a parental cell gives rise new daughter cells.
- **Cell wall:** In plant, a non-living part of cell covered and protects living cell organelles. It also helps to maintaining the structure of the cell and to hold on to water within the cell.

- **Cell:** A basic unit of any organism or individual responsible for various activities (structural and function). In other words, cell is the basic unit of life.
- **Cellulose:** Cellulose is an important structural component of the primary cell wall of green plants, many forms of algae and the oomycetes. The cellulose content of cotton fiber, wood and dried hemp is 90% and 40–50%, respectively.
- **Centriole:** It is a small cylindrical body about 0.4 micrometer long and about 0.15 micrometer in diameter, consisting mainly of nine parallel tubular structures arranged in the form of a cylinder. The two centrioles of each pair lie at right angles to each other.
- **Centrioles:** It is paired structures in the centrosome. Each centriole is comprised of nine triplet microtubules.
- **Centrosomes:** This is microtubule structure involved in movement of chromosomes (DNA) during cell division.
- **Chloroplasts:** Plants contain chloroplasts where photosynthesis occurs. Chloroplasts are organelles which are not present in animal cells. Photosynthesis is the conversion of CO_2 and water in the presence of sunlight into carbohydrates as a food source for the plant and O_2 as a by-product.
- **Chromatin:** DNA–protein complex.
- **Chromoplast:** Plastids that contain high concentrations of carotenoid pigments rather than chlorophyll.
- **Cilia:** A fine cellular folding of the cell membrane typically associated with the apical (top) surface of a cell responsible for absorption and motility.
- **Closed mitosis:** A process refers to separation of chromosomes into two daughter nuclei within nucleus.
- **Cristae:** The folds of the inner membrane form compartments present in mitochondria. In other words, cristae are ultramicroscopic features of mitochondria.
- **Crossing over:** A process, by which recombinant chromosomes are produced during meiosis.
- **Cytokinesis:** Process by which a cell is partitioned into two progeny cells. Cytokinesis usually begins late in mitosis.
- **Cytology:** The study of the structure and function of cells.
- **Cytoplasm:** The cytoplasm or cytosol is contained by the cell membrane. Cytoplasm is an archaic term based on primitive understanding of the cellular plasma or fluid.
- **Cytoplasmic streaming:** It is the coordinated movement of particles and organelles through the cytosol in a helical path down one side of a cell and up the other side.
- **Cytoskeleton:** Electron microscopy reveals several types of protein filaments criss crossing the eukaryotic cell, forming an interlocking three-dimensional meshwork.
- **Disomic:** Individual which contains pair of each chromosome of the genome.

- **Endocytosis:** In a multistep process, a pocket is formed in the cell membrane allowing some extracellular material to be engulfed.
- **Endo-mitosis:** A process by which the division of chromosomes occurs without cell division.
- **Endoplasmic reticulum (ER):** The mechanism of transport across the membrane is complex, involving the ribosomes, the mRNA that codes for the secretory protein, and a special receptor in the ER membrane.
- **Endosome:** A lysosome may fuse with the endocytosed vesicle to form an endosome.
- **Eukaryotes:** The structurally more complex eukaryotic cells include protists, fungi, plants, and animals.
- **Exocytosis:** A process by which cells excrete numerous products and waste materials from the cell. Exocytosis is especially important for polar or water-soluble materials which cannot freely pass through the cell membrane.
- **G_1 phase:** A sub phase of interphase in which, nuclear DNA is prepared for replication by the assembly of a pre-replication complex at the origin of replication along the chromatin.
- **Glucose:** A principal food compound of cells and breakdown, energy and "reducing power," both of which are essential in biosynthetic reactions are salvaged and stored, mainly in the form of ATP in the case of energy and NADH for reducing power.
- **Glyoxysomes:** It contains the *glyoxylate cycle*, which helps to convert stored fatty acids into sugars that can be translocated throughout the young plant to provide energy for growth.
- **Golgi apparatus:** The stacks of membranes that make up the golgi apparatus are located between the nucleus of the cell and the cell membrane in an orderly fashion. The Golgi apparatus plays a key role in the synthesis and secretion of complex polysaccharides (polymers composed of different types of sugars) and in the assembly of the oligosaccharide chains of glycoproteins.
- **Hayloplasm:** Materials of cytoplasm except other structure present in cell.
- **Hemi-cellulose:** Hemi-cellulose is a polysaccharide related to cellulose that comprises about 20% of the biomass of most plants.
- **Heterochromatin:** A highly compact and transcriptionally inactive form of chromatin.
- **Homeostasis:** Maintenance of nearly constant conditions in the environment.
- **Karyokinesis:** A process by which the division of chromosomes into daughter cell from pre-existing cell.
- **Keratins:** A type of intermediate filament found in the cytoplasm, are composed of a different polypeptide monomer.
- **Kinetochores:** They are located in the centromeric regions of the chromosomes. Some of the unattached microtubules overlap with microtubules the opposite polar region in the spindle mid-zone.

- **Leucoplasts:** Non-pigmented plastids or a starch storing plastid. Amyloplasts are abundant in storage tissues of the shoot and root, and in seeds.
- **Lysosome:** The term "some" means "body" and "lyse" means "destroy," so, a lysosome is a body which destroys.
- **Matrix:** The compartment enclosed by the inner membrane, the mitochondrial and contains the enzymes of the pathway of intermediary metabolism called the Krebs cycle.
- **Metaphase:** The chromatids are pulled tightly by their attached microtubules to the center of the cell, lining up to form the *equatorial plate* of the mitotic spindle.
- **Microfilaments:** The filaments are solid, with a diameter of 7 nm; they are composed of a special form of the protein found in muscle: globular actin, or G-actin. A microfilament consists of two chains of polymerized actin subunits that intertwine in a helical fashion.
- **Microscope:** An instrument that provides a magnified image of a tiny object.
- **Microtubules:** They are composed of repeating protein subunits called tubulin. Microtubules are also the functional component of flagella.
- **Microvilli:** Cell extensions filled with a network of cross-linked actin filaments.
- **Mitochondria:** Mitochondria are colloquially called the "powerhouse" of the cell and are the cellular sites of respiration, a process in which the energy released from sugar metabolism is used for the synthesis of ATP (adenosine triphosphate) from ADP (adenosine diphosphate) and inorganic phosphate (Pi).
- **Mitosis:** The process by which previously replicated chromosomes are aligned, separated, and distributed in an orderly fashion to daughter cells.
- **Multicellular:** Individual formed by multiple cells.
- **Myosins:** The proteins that have the ability to hydrolyze ATP to ADP and Pi when activated by binding to an actin microfilament. Myosins belong to the general class of **motor proteins** that drive cytoplasmic streaming and the movements of organelles within the cell.
- **Nuclear lamins:** In animal cells, a specific polypeptide is composed of monomer.
- **Nucleoids:** A DNA circles are localized in specific regions of the mitochondrial matrix or plastid stroma.
- **Nucleolus:** Inside the nucleus is found the nucleolus, or little nucleus. This spherical mass is composed of RNA and proteins. The nucleolus is the source of the ribosomes.
- **Nucleolus:** Nuclei contain a densely granular region.
- **Nucleosome**: Segments of the linear double helix of DNA are coiled twice around a solid cylinder of eight histone protein molecules.
- **Nucleus:** A part of cell contains nucleic acid *i.e.*, RNA and DNA.
- **Pachytene:** A sub-phase of prophase I of meiosis responds to crossing over between homologous chromosomes.

- **Peroxisomes:** It is found in all eukaryotic organisms, and in plants, they are present in photosynthetic cells. Peroxisomes function both in the removal of hydrogen from organic substrates, consuming O_2 in the process.
- **Pinocytosis (cellular sipping):** Forming small endocytotic pockets.
- **Plasmodesmata:** Tubular extensions of the plasma membrane, 40 to 50 nm in diameter, that traverse the cell wall and connect the cytoplasms of adjacent cells.
- **Prokaryotes:** The structurally simpler molecules *i.e., bacteria.*
- **Prometaphase:** It is early stage of metaphase and multiple microtubules from the aster attach to the chromatids at the centromeres, where the paired chromatids are still bound to each other; the tubules then pull one chromatid of each pair toward one cellular pole and its partner toward the opposite pole.
- **Prophase spindle:** A structure form in prophase, the microtubules begin to assemble at two foci on opposite sides of the nucleus.
- **Prophase:** First stage of mitosis.
- **Protein:** Proteins are composed of long chains of amino acids linked by amide bonds, known as peptide bonds.
- **Quantasomes:** A group of grana lamelle and stroma found in chloroplast.
- **Rough ER:** It is also a part of E. R. and the site of synthesis of membrane proteins and proteins to be secreted outside the cell or into the vacuoles.
- **Secondary cell walls:** It is thicker and stronger than primary walls and are deposited when most cell enlargement has ended. Secondary cell walls owe their strength and toughness to lignin, a brittle, glue like material.
- **Secretory vesicles:** Vesicles derived from the Golgi carry the polysaccharides and glycoproteins to the plasma membrane, where the vesicles fuse with the plasma membrane and empty their contents into the region of the cell wall. Secretory vesicles may either be smooth or have a protein coat.
- **Smooth ER:** A type of E.R has functions as a major site of lipid synthesis and membrane assembly.
- **Telophase:** In telophase, the two sets of daughter chromosomes are pushed completely apart.
- **Topoisomerase I:** An enzyme responsible for super coiling of chromosomes condensing.
- **Tubulin:** A monomer of microtubules is a hetero-dimer composed of two similar polypeptide chains (α- and β-tubulin), each having an apparent molecular mass of 55,000 daltons.
- **Unicellular:** Individual formed by single cell *i.e.*, bacteria, algae, etc.

ANSWERS

Chapter 1: Cell and Cell Division

1. C	41. A	81. B	121. B	161. D	201. C
2. C	42. C	82. A	122. A	162. A	202. A
3. A	43. A	83. A	123. A	163. A	203. C
4. A	44. B	84. A	124. B	164. C	204. C
5. A	45. B	85. C	125. A	165. B	205. A
6. D	46. C	86. B	126. C	166. D	206. A
7. D	47. B	87. A	127. B	167. D	
8. A	48. A	88. A	128. A	168. A	
9. A	49. A	89. C	129. A	169. D	
10. A	50. B	90. A	130. D	170. D	
11. B	51. A	91. D	131. A	171. A	
12. D	52. D	92. A	132. D	172. A	
13. A	53. A	93. C	133. B	173. D	
14. A	54. B	94. A	134. D	174. D	
15. B	55. C	95. A	135. B	175. D	
16. A	56. C	96. A	136. D	176. A	
17. A	57. A	97. A	137. A	177. A	
18. A	58. C	98. C	138. B	178. B	
19. A	59. A	99. C	139. B	179. B	
20. A	60. C	100. D	140. A	180. D	
21. B	61. A	101. A	141. D	181. A	
22. B	62. A	102. A	142. A	182. D	
23. C	63. B	103. A	143. C	183. D	
24. A	64. A	104. D	144. A	184. D	
25. A	65. D	105. C	145. B	185. D	
26. B	66. D	106. C	146. D	186. C	
27. B	67. A	107. A	147. D	187. A	
28. A	68. B	108. B	148. D	188. A	
29. B	69. A	109. A	149. D	189. A	
30. C	70. B	110. A	150. B	190. D	
31. D	71. A	111. C	151. D	191. A	
32. C	72. B	112. B	152. B	192. C	
33. A	73. B	113. B	153. A	193. D	
34. A	74. A	114. A	154. A	194. A	
35. A	75. B	115. C	155. D	195. A	
36. C	76. C	116. D	156. D	196. A	
37. C	77. D	117. D	157. A	197. B	
38. C	78. B	118. C	158. D	198. A	
39. C	79. A	119. D	159. B	199. A	
40. A	80. C	120. A	160. A	200. A	

Chapter 2

Chromosomes

1. Chromosome was discovered by :
   A. Waldeyer in 1875
   B. E. Strasburger, 1875
   C. Both A & B
   D. None of these
2. Who coined the term chromosome?
   A. Waldeyer, 1888
   B. Strasburger, 1888
   C. Negli, 1888
   D. None of these
3. Chromatin is composed of :
   A. Protein
   B. DNA
   C. RNA
   D. All of these
4. W. S. Sutton and T. Boveri were associated with which of the following event in 1903:
   A. Law of segregation
   B. Chromosomal theory of inheritance
   C. Both A & B
   D. None
5. The homologous chromosomes become separate in which of the following event?
   A. Mitosis
   B. Meiosis
   C. Both A & B
   D. None
6. A rod like and thickest stature appear in metaphase of cell division is known as :
   A. Chromatin
   B. Chromosome
   C. Both A & B
   D. None of these
7. Which of the following component is found in chromatin?
   A. DNA
   B. RNA
   C. Protein
   D. All of these
8. Which of the following shape of metacentric chromosomes appears in anaphase in cell division:
   A. V shape
   B. J shape
   C. Both A & B
   D. None of these

9. At interphase, a thin thread like microscopic visible stature seen is known as :
   A. Chromatin B. Chromosome
   C. DNA D. None of these
10. In which phase duplication of chromosomes takes place:
   A. Interphase B. Prophase
   C. Metaphase D. None of these
11. The basic chromosome set of a species is known as ________ of the same species.
   A. Gene B. Genome
   C. Genomics D. Haploid
12. Which of the following symbol shows basic chromosome in cells?
   A. X B. n
   C. 2n D. All of these
13. Which of the following symbol shows gametic chromosome numbers?
   A. 2X B. n
   C. 2n D. All of these
14. Which of the following symbol shows somatic chromosomes of cells?
   A. 2n B. n
   C. X D. All of these
15. Which of the following symbol is used to show true diploid species?
   A. 2x B. 2n = 2x
   C. Both A & B D. All of these
16. Basic chromosome number is also known as:
   A. Monoploid chromosome number B. Diploid chromosome number
   C. Polyploid chromosome number D. All of these
17. Which of the following cell contains somatic chromosomes?
   A. Germ cell B. Vegetative cell
   C. Gametic cell D. All of these
18. Which of the following cell contains gametic chromosomes?
   A. Germ cell B. Gametic cell
   C. Both A & B D. None of these
19. What are the homologous chromosomes?
   A. Two identical chromosomes in all aspects
   B. Two iso-chromosomes, present in pairing condition
   C. Both A & B
   D. A group of chromosomes that contain similarities except one gene

20. Suppose male and female gametes are present in a flower. Which of the following product will develop by the fusion of above gametes?

A. Diploid
B. Monoploid
C. Zygote
D. All of these

21. How many chromosome numbers are in growing cells of ant (*Myrmecia pilosula*)?

A. 2
B. 4
C. 6
D. 8

22. How many somatic chromosomes are present in *Haplopappus gracilis*?

A. 2
B. 4
C. 6
D. 8

23. Gametic chromosomes are also known as:

A. Haploid chromosomes
B. Somatic chromosomes
C. Both A & B
D. None of these

24. During cell division, a bead likes nodules appears in leptotene is known as:

A. Chromomeres
B. Centromeres
C. Telomere
D. All of these

25. Which of the following phase of cell division is suitable for measurement of chromosomes?

A. Mitotic metaphase
B. Meiotic metaphase
C. Both A & B
D. None of these

26. Which of the following phase of cell division is suitable for observing the shape of chromosome?

A. Anaphase
B. Telophase
C. Metaphase
D. Prophase

27. A point at which sister chromatids of a chromosome are linked is called :

A. Centromeres
B. Primary constriction
C. Both A & B
D. None of these

28. Which of the following chromosomes have satellite type structure?

A. SAT chromosomes
B. Normal chromosome
C. Both A & B
D. None of these

29. How many animal species carry supernumerary or extra chromosomes?

A. More than 100 species
B. 75 species
C. 50 species
D. None

30. Within chromosome, a coiled filamentous structure is known as :

A. Chromonema
B. Secondary construction
C. Primary construction
D. All of these

31. A beaded structure present on chromonema is also called:

A. Gene bearing portion
B. Centromeres
C. Telomeres
D. All of these

32. A chromosome that has its two terminal portions is known as :

A. Centromeres B. Telomeres

C. NOR D. None of these

33. Which of the function is played by telomeres?

A. Do not allow any mixing of chromosomes through telomere region

B. Preventing the chromosomes shortening

C. Both A & B

D. None of these

34. Satellite DNA is located at :

A. Telomeric region B. Centromeric region

C. Both A & B D. None of these

35. The complete selection of chromosomes in an organism is termed as:

A. Karyotype B. Cytokinesis

C. Both A & B D. None of these

36. Removal of male bud in banana is technically called as :

A. Tipping B. Disbudding

C. Denovelling D. All of these

37. Who observed chromosomes in plant cell in 1842?

A. Karl Wilhelm von Nägeli B. Walther Flemming

C. Von Waldeyer D. None of these

38. Who described the behavior of chromosomes in animal cell during mitosis in 1882?

A. Karl Wilhelm von Nägeli B. Walther Flemming

C. von Waldeyer D. None of these

39. Chromosomes consists of:

A. DNA B. RNA

C. Histone proteins and non histone protein D. All of these

40. Karyotype represents which of the following chromosomes :

A. Gametic chromosomes B. Somatic chromosomes

C. Basic chromosomes D. Both A & C

41. Study of a chromosomal set of an organism or species is known as :

A. Karyology B. Cytology

C. Genetics D. Both A & B

42. Which of the followings, technique is used for preparation of karyotypes of chromosomes based on differentiating the morphology of chromosomes within a species?

A. Chromosome banding technique B. FISH technique

C. Flow cytometry D. All of these

43. Which of the plant karyotype is prepared by using the flow cytometry?

A. Tomato
B. Faba bean
C. Wheat
D. All of these

44. Which of the following regions is detected on chromosomes by using banding technique?

A. GC rich region or guanine-cytosine rich region
B. AT rich region or adenine – thymine rich region
C. Both A & B
D. None of these

45. Which of the followings banding technique is used in plant system for detection of karyotypes?

A. Giemsa C banding technique
B. Giemsa N banding technique
C. Giemsa Q banding technique
D. Both A & B

46. Which of the following features is found in asymmetric Karyotype?

A. A large difference between smallest and largest chromosome
B. A group of chromosomes show similarities between them
C. Both A & B
D. None of these

47. Symmetrical karyotype is mostly used to find out which of the following:

A. Primitive species
B. Allopolyploidy
C. True diploids
D. Both A & C

48. Karyology refers to:

A. Study of chromosomes
B. Study of nucleus
C. Study of cell and its structural part
D. All of these

49. The study of complete sets of chromosomes, is also known as:

A. Karyology
B. Cytology
C. Both A & B
D. None of these

50. Karyology is a part of :

A. Cytology
B. Genetics
C. Molecular genetics
D. All of these

51. Karyotypes consisted of which of the following class:

A. Karyogram
B. Ideogram
C. Both A & B
D. None of these

52. Which of the the following dye is used for staining the chromosomes for detection of their karyotypes:

A. Hoechst 33258
B. Olivomycin
C. Giemsa
D. All of these

53. Hoechst 33258 is a dye used for staining which of the following region of chromosomal segments in human:
    A. A-T rich region
    B. G-C rich region
    C. Both A & B
    D. None of these
54. Giemsa dye is used to stained which one of the following region of the chromosomes in human is:
    A. G-C region
    B. A-T region
    C. Both A & B
    D. None of these
55. Olivomycin dye is used for staining which of the following region of chromosomes in human:
    A. A-T rich region
    B. A-T light region
    C. G-C rich region
    D. Both A & B
56. Symmetrical karyotype represents :
    A. Same size of all metacentric chromosome
    B. Dissimilarities in size of chromosomes
    C. Chromosome in short and not similar in shape
    D. All of these
57. Asymmetrical karyotypes of chromosome represents:
    A. Similar chromosome in metaphase
    B. Deviation in size of all metacentric chromosome
    C. No change in gene constitution of chromosome
    D. All of these
58. Identical karyotypes found in :
    A. All members of same species
    B. All members of different genera
    C. All genera of same family
    D. Same family
59. Morphological features of chromosome represented by a diagram is known as:
    A. Ideogram
    B. Idiotype
    C. Heterochromatin
    D. All of them
60. Idiotype is prepared for detection of :
    A. Somatic chromosome of a species
    B. Haploid chromosome complement of a species
    C. Diploid chromosomes of a species
    D. All of these
61. Chromosome consist of :
    A. Chromatin
    B. Gene
    C. Protein
    D. All of these
62. Chromosome binding technique was developed for :
    A. Analysis of heterochromatin extension in chromosome
    B. Distribution of gene in cell
    C. Assessment of frequency of gene
    D. Study of mutation

63. Constitutive heterochromatin is consisted of genetical :

A. Active fluid | B. Inactive material
C. Dynamic material | D. Vigorous fluid

64. Gametes of a species contain :

A. Full amount of DNA | B. ½ amount of DNA of a somatic cell
C. 1/3 amount of DNA of a somatic cell | D. 1/5 amount of DNA of a somatic cell

65. Which of the following DNA is found in chromosome?

A. Repetitive DNA | B. Unique DNA
C. Both A & B | D. None

66. Unique DNA consist of :

A. Single copy per genome | B. Double copy per genome
C. Linear sequence of DNA | D. Only A & C

67. Protein associated with chromosome is :

A. Basic protein | B. Histone protein
C. Amino protein | D. Only A & B

68. What percentage of unique DNA is found in *Drosophilla melanogaster?*

A. 70% | B. 80%
C. 50% | D. 30%

69. What percentage of unique DNA is found in Maize ?

A. 24% | B. 40%
C. 22% | D. 70%.

70. Nature of Non histone protein, which is found in chromosome, is :

A. Acidic | B. Base
C. Saline | D. None

71. What is the ratio of histone protein in total chromosomal protein?

A. 60% | B. 50%
C. 20% | D. 80%

72. One μ of human meta-centric chromosome length equals one of the following:

A. Approx. 0.0255 pg | B. Approx. 3.008 pg
C. Approx. 10.99 pg | D. None

73. What percent of repetitive DNA is found in rye (*Secale cereale* L.)?

A. 83% | B. 92%
C. 50% | D. 20%

74. Ratio of repetitive DNA which is found in bread wheat (*Triticum aestivum* L.)?

A. 83% | B. 23%
C. 70% | D. 40%

75. Proportion of repetitive DNA that is found in barley (*Hordeum vulgare* L.)?

A. 70% B. 76%

C. 90% D. 34%

76. RNA associated with chromosome is of one of the following:

A. rRNA B. mRNA

C. tRNA D. All of these

77. How many DNA double helix contain one μ of a metaphase chromosome?

A. 1500 μ B. 8000 μ

C. 4000 μ D. 2000 μ

78. Size of linker DNA is Nucleosome manage between?

A. 8 bp to 114 bp B. 1 bp to 3 bp

C. 30 bp to 37 bp D. 90 bp to 100 bp

79. Which of the following, histone protein is mostly found at linker DNA and scals the DNA in the nucleosome?

A. H1 B. H2A

C. H2B D. H3

80. Nucleosome fibre is also called:

A. 13nm fibre B. 10nm fibre

C. 100nm fibre D. 80nm fibre

81. Who demonstrated the replication of chromosomes in 1956?

A. Taylor, J.H. B. Johnson

C. Jensen D. None

82. Special chromosomes are:

A. Lampbrush chromosome B. Polytene chromosome

C. B chromosome D. All of these

83. Which of the following term is related to lampbrush chromosome?

A. Loop B. Puff

C. Balbiyani ring D. None

84. Which of the following enzyme is able to digest the loop of lampbrush chromosome?

A. RNase B. Protease

C. Both A & B D. Hydrosylase

85. In which site genes are located in lampbrush chromosome?

A. Puff B. Knob

C. Loop D. All of these

86. Giant chromosome is found in:

A. Salivary glands of larvae B. Gut epithelium

C. Malpighian D. All of these

87. Who reported polytene or giant chromosome in salivary glands of dipteran in 1881?

A. Balbiyani B. Muller

C. Morgan D. None

88. Formation of chromocentre in polytene chromosome through which of the following process:

A. Fusion of the centromeric region of all chromosome

B. Fusion of the terminal region of all chromosome

C. Separation of all centromeric region of all chromosome

D. None

89. Which of the following term is related with polytene chromosome?

A. Loop B. Puffs

C. Balbiani ring D. Only B & C

90. Which of the following site is responsible for active RNA synthesis in polytene chromosome?

A. Puffs B. Terminal region

C. Both A & B D. Matrix

91. B-chromosome is also known as:

A. Accessory chromosome B. Special chromosome

C. Supernumerary chromosome D. All of these

92. B-chromosome is similar with somatic chromosome in their:

A. Morphology B. Function

C. Genetic constitution D. All of these

93. Who suggested the role of chromosome in heredity?

A. Mendel and Bateson B. Koelreuter and Knight

C. Sutton and Boveri D. None

Terms

- **Basic chromosome number:** It is also known as monoploid. Basic chromosome refers to the number of half chromosomes of true diploid species. It's designated by x.
- **Centromeres or primary construction:** A portion of chromosome found between two chromatids.
- **Chromatid interference:** A condition in which the occurrence of a cross over between any two non sister chromatids can be shown interfere with the same chromatids being involved in other crossover in the same meiosis.
- **Chromatin:** The deoxyribonucleohistone in chromosome.

- **Chromocentre:** The point at which polytene chromosome appear to be attached.
- **Chromomeres:** Bread like nodules seen on chromosome.
- **Chromosome banding:** By straining of chromosome light and dark region occur along the length of chromosomes.
- **Chromosome marker:** The markeable feature of chromosome can be used for evolution.
- **Chromosome remodeling:** The change in chromatin structure due to DNA histone interaction in the nucleosome, to facilitate interaction of other factors with DNA template for transcription.
- **Chromosome:** A thread like body appears during cell division. In other hands, darkly straining nucleoprotein bodies that observed in cell division.
- **Diploid:** Cell having two sets of chromosomes (2n).
- **Euchromatin:** Chromatin region is transcriptionally active and content most of the protein coating gene.
- **Gametic chromosome:** Chromosome found in gametic cell.
- **Heterochromatin:** Chromatin contains un-active genes.
- **Homologous chromosome:** Two identical chromosomes.
- **Ideotype:** A technique by which morphological feature of chromosomes is represented.
- **Karyology:** Study of chromosome.
- **Polytene chromosome:** A special type of chromosome found in salivary glands of dipteran.
- **Puffs:** A site of chromosome responsible for active RNA synthesis in polytene chromosome.
- **SAT chromosome:** Chromosome has satellite type structure.
- **Telomere:** Terminal portion of chromosome.
- **Histone:** A group of protein rich in basic amino acids.
- **Chromosome substitution:** Replacement of a pair of chromosome of one strain by another.
- **Chromosome walking:** It is used in identifying overlapping DNA fragments and used in preparation of restriction maps of small regions of chromosome.
- **Acentric chromosomes:** Chromosome without centromere.
- **Acrocentric chromosomes:** Chromosome has centromere at their terminal region.
- **Amphidiploids:** An allopolyploid has two copies of each of two or more genomes; behaves like diploid during meiosis.
- **Colchicine:** An alkaloid to be used for chromosome doubling.
- **Pellecile:** A membrane which covers each chromosomes
- **Submetacentric chromosome:** L or J shape chromosomes.
- **Dicentric chromosome:** Chromosome having two centromeres

ANSWERS

Chapter 2: Chromosome

1. B	16. A	31. A	46. A	61. D	76. D	91. D
2. A	17. B	32. B	47. D	62. A	77. B	92. A
3. D	18. C	33. C	48. A	63. B	78. A	93. C
4. B	19. A	34. A	49. A	64. B	79. A	
5. B	20. C	35. A	50. A	65. C	80. B	
6. B	21. A	36. C	51. C	66. D	81. A	
7. D	22. A	37. A	52. D	67. D	82. D	
8. A	23. A	38. B	53. B	68. A	83. A	
9. A	24. A	39. D	54 A	69. C	84. C	
10. A	25. A	40. D	55. A	70. A	85. C	
11. B	26. A	41. D	56. A	71. D	86. D	
12. A	27. C	42. D	57. B	72. A	87. A	
13. B	28. A	43. D	58. A	73. B	88. A	
14. A	29. A	44. C	59. B	74. A	89. D	
15. C	30. A	45. D	60. B	75. B	90. A	

Chapter **3**

Crossing over and Linkage

1. A chromosome has number of genes and some of them do not separate from each other in:

   A. Mitosis
   B. Meiosis
   C. Epigenetic
   D. All of these

2. A chromosome has many genes and some of the genes go together in the next generation as they were in parental chromosome, this phenomenon is known as:

   A. Linkage
   B. Independent assortment of genes
   C. Single gene inheritance
   D. Poly genes inheritance

3. Which terms are related to linkage?

   A. Coupling phase
   B. Repulsion phase
   C. Both A & B
   D. None

4. How many pairs of chromosomes are found in *Pisum sativum?*

   A. 6
   B. 8
   C. 14
   D. 7

5. In pea, the genes controlling the expression of seed color and seed shape are found :

   A. Chromosome no. 1 and chromosome no. 7
   B. Chromosome no. 3 and chromosome no. 4
   C. Only chromosome no. 1
   D. Only single chromosome no. 7

6. Name of genes which do not get separated from each other during crossing over :

   A. Linked gene
   B. Linkage
   C. Both A & B
   D. None

7. Which cross deals with the separation of two pairs of alleles independently ?

   A. Poly cross
   B. Single cross
   C. Dihybrid cross
   D. Monohybrid cross

8. Question no. 7. Follows the law of inheritance:
   A. Law of independent assortment B. Law of segregation
   C. Both A & B D. None of these
9. Linkage represents :
   A. Failure of gene pairs to assort independently
   B. Relationship between different pairs of genes
   C. Both A & B
   D. None
10. Who proposed the relationship between Mendelian factor and chromosome in 1903?
   A. G. J. Mendel B. T. H. Morgan
   C. Sutton D. None
11. Exchange of chromosomal segments take place in:
   A. Crossing over B. Translocation
   C. Both A & B D. None
12. In case of linkage, the distance between two or more genes is :
   A. Less distance B. More distance occurs
   C. Chromosomal segment absent D. None
13. Deviation in double gene inheritance of Mendel will be reflected as the change in 1:1:1:1 test cross ratio due to:
   A. Linkage B. Epistasis
   C. Both A & B D. None
14. Who proposed the hypothesis of coupling and repulsion phases?
   A. Bateson B. Punnett
   C. Darwin D. None
15. Which condition arises due to independent assortment of genes in a dihybrid test cross progeny ?
   A. 1:1:1:1 B. 1:2:3:1
   C. 4:2 D. All of these
16. Which condition arises due to linkage between dominant and recessive alleles or vice-versa?
   A. Repulsion phase of linkage B. Coupling phase of linkage
   C. Both A & B D. None
17. Transmission of genes in group is known as:
   A. Heredity B. Linkage
   C. Crossing over D. Synapsis
18. Tendency of genes to be inherited in a group was observed in pea in 1906 by:
   A. W. Bateson B. R. C. Punnett
   C. Both A & B D. None

19. In linkage, which event occurs?
    A. Segregation of alleles
    B. Non segregation of alleles
    C. Crossing over
    D. None
20. On linkage, Morgan published his paper in:
    A. 1911
    B. 1955
    C. 1901
    D. 1900
21. In Drosophilla, a cross between homozygous flies with long wings and grey bodies and homozygous flies with short wings and black bodies. Which results will be produce in F_1?
    A. All F_1 flies have long wings and gray bodies
    B. All F_1 flies have short wings and gray bodies
    C. All F_1 flies have long wings and black bodies
    D. None
22. The arrangement of mutant form with normal form of gene is called :
    A. Coupling
    B. Repulsion
    C. Cis configuration
    D. Both A & C
23. Which process is known as interchange of gene between homologous chromosome ?
    A. Crossing over
    B. Linkage
    C. Translocation
    D. Inversion
24. Which process is important for evolution ?
    A. Crossing over
    B. Inversion
    C. Nullisomy
    D. All of these
25. Which process produces new combination of genes ?
    A. Gametes of purity
    B. Independent assortment
    C. Crossing over
    D. All of these
26. Natural selection represents:
    A. Goodness of genotypes
    B. Good Fitness
    C. Both A & B
    D. Homeostasis
27. Which event results in crossing over ?
    A. Loci arranged in linear sequence
    B. Breakage of allele
    C. Exchange of genes
    D. All of these
28. Synapsis event occurs in:
    A. Mitosis phase
    B. Inter-phase
    C. Prophase I
    D. None
29. Which event responds to new combination of alleles between sisters chromatids of homologous chromosome?
    A. Translocation
    B. Termination
    C. Duplication
    D. Cross over

30. Exchange of genes between non sister chromatids of non homologous chromosome:
   A. Translocation B. Termination
   C. Duplication D. Cross over
31. How many chromatids are in each pairs of chromosomes?
   A. Six B. Four
   C. Two D. All of these
32. Which of the following phase carries synapsis that is required for crossing over?
   A. Prophase I B. Prophase II
   C. Mitosis D. Inter-phase
33. Which condition is helpful for occurence of high crossing over frequency between two loci ?
   A. Less distance between two loci on the chromosome
   B. More distance between two loci on the chromosome
   C. No distance between two loci on the chromosome
   D. None
34. Minimum genes or segments, where each pair of homologous chromosomes will be required for crossing over :
   A. Two B. Three
   C. Four D. Six
35. Who observed cross shape configuration of chromosomes for exchange of gene/ genes between a pair of homologous chromosome in amphibians ?
   A. F. Janssen B. Sutton
   C. Johnson D. Morgan
36. F. Janssen detected a cross shape use for exchange gene/ genes between a pair of homologous chromosome in amphibians is called :
   A. Chiasma B. Tetrad
   C. Holi D. None
37. Who put evidence of cytological basis of crossing over of homologous chromosomes working on *Drosophilla* ?
   A. C. Stern, 1931 B. C. Stern, 1980
   C. C. Stern, 1900 D. None
38. H.B. Creighton and B. Mc Clintock proposed cytological basis of crossing over for exchange of chromosomal segments of each pairs of homologous chromosomes in :
   A. Maize B. Pea
   C. Sugar beat D. None
39. Each pair of synapsed homologous chromosome is called :
   A. Tetrad B. Monod
   C. Chromatids D. None

40. Crossing over differs from translocation due to:
A. Occurrence between a pair of homologous chromosomes
B. Occurrence between a pair of non homologous chromosome
C. Both A & B
D. None

41. Which feature is most important for linkage maps?
A. Linearity B. Branches
C. Both A & B D. None

42. Linkage maps are circular in :
A. Viruses B. Prokaryotes
C. Eukaryotes (mitochondria and chloroplast)
D. All of these

43. One linkage map is equal to :
A. One per cent recombinant chromosome B. One per cent recombinant gametes
C. Both A & B D. None

44. 1cM stands for :
A. One centi Morgan B. Morgan
C. One Celsius unit D. None

45. How many linkage groups are possible in *Drosophila melanogaster?*
A. 4 B. 5
C. 8 D. 2

46. Maximum linkage groups found in maize are:
A. 12 B. 20
C. 10 D. 8

47. How many pairs of chromosomes are found in maize?
A. 10 B. 20
C. 12 D. None

48. Who developed mapping function in 1919?
A. Haldane, J.B.S. B. Morgan, T. H.
C. Janssens, F. D. Khurana, K.

49. Mapping function is also known as:
A. Haldane function B. Holiday junction
C. Tetrad D. Cross over

50. The maximum frequency of recombination observed between genes located on the same chromosome is:
A. 50% B. 60%
C. 20% D. None

51. Who made somatic hybridization in mouse cells in 1960 ?
A. Barshki, G
B. Morgan, T. H.
C. Both A & B
D. None

52. Cell fusion will produce:
A. Binucleate hybrid cell
B. Heterokaryons cell
C. Synkaryons cell
D. All of these

53. Which product will be produced when the two somatic cells of the same species are fused ?
A. Mono-specific hybrid
B. Di-specific hybrids
C. Inter-specific hybrids
D. All of these

54. When mouse and human somatic cells are fused which features will be visible ?
A. Human and mouse chromosome are distinguishable
B. All the chromosomes are represented in the hybrid
C. Hybrid cell will have single chromosome of each genome
D. All of these

55. Crossing over is associated with a set of stature, which forms during prophase I is known as:
A. Synaptinemal complex
B. Chromosome
C. Cell
D. None

56. Synaptinemal complex is composed of :
A. Protein
B. DNA
C. RNA
D. Both A & C

57. Male *Drosophilla* does not show which of the following ?
A. Synaptinemal complex
B. Crossing over
C. Both A & B
D. None

58. Which enzyme is used for cleaved strands to join in recombinant arrangement:
A. DNA Ligase
B. DNA helicase
C. Both A & B
D. DNA unwinding protein

59. Double strand break model of crossing over was proposed in 1983 by:
A. Szostak *et al.*
B. Bolgord
C. Nermac
D. None

60. Which protein is most important for recombination between homologous chromosome as well as recombination of homologous DNA molecules?
A. Rec A
B. Protein A
C. Protein B
D. None

61. Illegitimate refers to :
A. Recombination between identical chromosome in all aspect
B. Recombination between non homologous chromosomes
C. Exchange of chromosomal segment with sister chromatids
D. None

62. The relationship between crossing over and formation of chiasma are based:-

A. Two plane theory
B. One plane theory or chiasma type theory
C. Both A & B
D. None

63. Who proposed two plane theory of crossing over in 1934?

A. Sharp
B. Creigton
C. Belling
D. Darlington

64. Two Plane Theory of crossing over is also known as:

A. Classical theory
B. One plane theory
C. Chismatype theory
D. All of them

65. Who proposed two plane theory of crossing over in 1909?

A. Janssen
B. Hutchinson
C. Sharp
D. Kolatker

66. Which theory has historical importance of crossing over?

A. One plane theory
B. Two plane theory
C. Both A & B
D. Chiasmatype theory

67. Who used copy choice model to explain the recombination of microbial system in 1955?

A. Janssen
B. Joshua Lederberg
C. Sharp
D. Kolatker

68. How many forms of a gene are carried by diploid eukaryotic cell?

A. Two
B. Three
C. Four
D. None

69. The structural results of a cross-strand exchange is known as:

A. Holliday junction
B. Holi junction
C. Both A & B
D. None

70. Crossing over occurs in a condition which arises due to:

A. Arrangement of aligned chromatids in meiosis
B. Leading to exchange of none homologous segments by homologous recombination.
C. Mutual exchange of chromosomal segment between non related chromosomes in prophase I
D. None

71. Who proposed the model of Holliday junction in 1964?

A. Robin Holliday
B. Naina Holi
C. Camillo Holi
D. None

72. The Holliday junction is a:

A. Tetrahedral structure
B. Dihedral structure
C. Monohedral structure
D. None

73. A new array of maternal and paternal alleles on the same chromosome is called :
   A. Recombination B. Mutation
   C. Both A & B D. None

74. Who demonstrated physical basis of crossing over in 1931?
   A. Harriet Creighton and Barbara McClintock
   B. C. Stern
   C. Fleming
   D. Muller

75. What happens when the genes get inherited together on the same chromosome?
   A. No change in character of an individual
   B. No change in genotype/ alleles combination
   C. Alleles show tightly linkup to each other
   D. All of these

76. Which process shows less change in linked genes?
   A. Linkage disequilibrium B. Migration
   C. Translocation D. Duplication

77. Non homologous cross over occurs :
   A. At the time of drastic mutation B. At the time of chromosomal aberration
   C. Both A & B D. None

78. Precocity theory was proposed by :
   A. Darlington, C. D. B. Robin Holliday
   C. Naina Holi D. Camillo Holi

79. Which factor affects the frequency of recombination?
   A. Distance between genes, sex, and age of female
   B. Temperature, nutrition, chemical and radiation
   C. Plasmagenes, genotypes, distance from centromere
   D. All of these

80. Expected ratio between the frequencies of chiasma and crossing over is:
   A. 1:1 B. 1:2
   C. 4:9 D. None

81. Which theory of crossing over is universally accepted?
   A. Chiasma type theory of crossing over B. Holliday junction
   C. Holi junction D. All of these

Terms

- **Chi square test:** A test to be used for testing the goodness of fit of observed ratio to expected ratio.
- **Chiasma:** A point representing place of segment an exchange between homologous chromosomes.
- **cM:** A map unit as referred to centiMorgan.
- **Coupling phase:** The phase refers to arrangement of mutant's alleles or normal alleles of two or more than two genes (AB/ab).
- **Crossing over:** Event responsible to produce new combination of alleles due to exchange of segments between sisters chromatids of homologous chromosome.
- **Frequency of recombination:** The proportion of recombinant genes between two gene pairs as compared to over all genes combination.
- **Holliday junction:** Structural results of a cross-strand exchange.
- **Illegitimate:** It refers to recombination between non homologous chromosomes.
- **Linkage disequilibrium:** Equation or process exhibits the differences between coupling and repulsion phase of linkage.
- **Linkage group:** The liked genes showing no independent assortment are grouped.
- **Linkage:** A chromosome has many genes and some of the genes go together in the next generation as they were in parental chromosome. In other hands, Transmission of genes in group is known as linkage.
- **Repulsion:** The arrangement of alleles of two genes in heterozygotes form *i.e.*, Ab/aB.
- **Synapsis:** Pairing of homologous chromosome during zygotene of prophase I in meiosis.
- **Synaptinemal complex:** Crossing over is associated with a set of stature which, forms during prophase I. Synaptinemal complex madeup by protein and RNAs.
- **Testcross:** Mating between heterozygotes and its recessive parents.
- **Tetrad analysis:** A technique is used for each of the individual spores of a tetrad for analysis to derive information on linkage and recombination.
- **Tetrad:** A picture made up by paired homologous chromosome during meiosis.
- **Translocation:** Event responds to new combination of alleles between non-homologous chromosomes.

ANSWERS

Chapter 3: Crossing over and linkage

1. B	16. A	31. B	46. C	61. B	76. A
2. A	17. B	32. A	47. A	62. C	77. C
3. C	18. C	33. B	48. A	63. A	78. A
4. D	19. B	34. A	49. A	64. A	79. D
5. A	20. A	35. A	50. A	65. A	80. A
6. A	21. A	36. A	51. A	66. A	81. A
7. C	22. D	37. A	52. D	67. B	
8. A	23. A	38. A	53. D	68. A	
9. C	24. D	39. A	54. D	69. A	
10. C	25. D	40. A	55. A	70. A	
11. C	26. C	41. A	56. D	71. A	
12. A	27. D	42. D	57. C	72. A	
13. A	28. C	43. C	58. A	73. A	
14. A	29. D	44. A	59. A	74. A	
15. A	30. A	45. A	60. A	75. D	

Chapter **4**

Law of Inheritance

1. Crossing refers to :

   A. Artificial pollination B. Natural pollination

   C. Both A & B D. None of these

2. The first generation obtaining raised by :

   A. 1st year generation B. F_1 progeny

   C. Both A & B D. None of these

3. F_1 stands for :

   A. First filial generation B. Second filial generation

   C. Synthetic generation D. None of these

4. Crossing may be one of the following types:

   A. Monohybrid crossing B. Dihybrid crossing

   C. Polyhybrid crossing D. All of these

5. Segregation refers to :

   A. The separation of two or more alleles of genes from each other.

   B. Separation of alternative phases of a gene from each other

   C. Both A & B

   D. None of these

6. Who was G. J. Mendel ?

   A. Father of Genetics B. Father of mutation

   C. Father of soil chemistry D. None of these

7. Which of the following crops was choosen by Mendel for his work?

   A. *Pisum sativum* L. B. *Cicer arietinum* L.

   C. *Triticum dicoccum* L. D. *Triticum aestivum* L.

8. Why did Mendel select pea (garden pea) for his working?
   A. Having contrast in forms of traits
   B. Having similar forms of characters
   C. Easily available
   D. Both A & C
9. What is direct cross?
   A. Cross between two individuals or one way cross
   B. Cross between same plant
   C. Both A & B
   D. None of these
10. Which of the following law of Mendel is universally accepted ?
    A. Law of gamete purity
    B. Law of independent assortment
    C. Both A & B
    D. None of these
11. Law of single segregation :
    A. Law of gamete purity
    B. Law of universally acceptance
    C. Law of applicable in metric traits inheritance
    D. All of these
12. Number of contrasting forms of each character was found in Mendel experiments ?
    A. Two
    B. Three
    C. Four
    D. None of these
13. Mendel used which of the term responsible for each character expression ?
    A. DNA
    B. Character
    C. Factor
    D. Gene
14. After modernization of biological science, an inherited factor responsible to determine a trait is known as:
    A. Gene
    B. DNA
    C. Both A & B
    D. None of these
15. Suppose, plant having tall stature and another plant showed dwarf phenotype, these traits governed by TT and tt types of allele of genes. A scientist made a cross between both tall and dwarf plant and obtained tall phenotype in F_1 generation. After F_1 seeds are selfed and obtained 1TT: 2Tt: 1tt types of genotypic expression and phenotypic ratio 3tall:1dwarf. Which of the following inheritance pattern involved above crossing product.
    A. Single gene inheritance pattern
    B. Double gene inheritance pattern
    C. Poly gene inheritance pattern
    D. None of these
16. Genotype refers to:
    A. Genetic makeup of an individual
    B. Phenotypic makeup of an individual
    C. Both A & B
    D. None of these

17. The sum of all biochemical, structural, molecular and physiological activities in the form of visibility is:
    A. Genotype
    B. Phenotype
    C. Both A & B
    D. None of these

18. Phenotype refers to:
    A. Visible expression of plant or organism
    B. Genetic constitution of plant or organism
    C. Both A & B
    D. None of these

19. Genetic variation refers to:
    A. Variation in genetic constitution of plants or organisms
    B. Variation in phenotypic expression of plant or organism
    C. Both A & B
    D. All of above

20. Homozygote refers as :
    A. Individual having different alleles of genes
    B. Individuals having identical alleles of genes
    C. Both A & B
    D. None of these

21. A condition, when an organism has two different alleles of a gene/genes is known as:
    A. Heterozygote
    B. Homozygote
    C. Both A & B
    D. None of these

22. Which of the following symbol is used to represent a recessive homozygous individual?
    A. Tt
    B. Aa
    C. AA
    D. aa

23. Which of the following symbol is used to represent a dominant homozygous individual?
    A. TT
    B. Tt
    C. Aa
    D. All of these

24. Which of the following sentence is correct according to law of segregation?
    A. Alleles of a gene do not contaminate each other
    B. Alleles of a gene is affected by each other
    C. Both A & B
    D. None of above

25. Alleles of a gene segregate during:
    A. Meiosis
    B. Mitosis
    C. Both A & B
    D. None of above

26. Alleles or genes segregate from each other in true or pure form according to:
    A. Law of segregation
    B. Law of co-domoinance
    C. Both A & B
    D. None

27. According to question no. 11, which of the cross is involved:
    A. Dihybrid cross
    B. Monohybrid cross
    C. Poly hybrid cross
    D. None of these

28. A Cross which is made for one character or one gene difference between two parents:
    A. Monohybrid cross
    B. Dihybrid cross
    C. Polyhybrid cross
    D. All of these

29. A cross was made between two distinguished plant A and B having red flower (RR) and white flower (rr), respectively. The F_1 progeny showed red flowered plants. Which of the following condition is responsible for the expression of flower color?
    A. Co-dominance
    B. Dominance
    C. Incomplete or partial dominance
    D. All of these

30. A cross was made between two dissimilar individuals and their F_1 showed the combination of both parental expression according to one of the following conditions:
    A. Dominance condition
    B. Partial dominance condition
    C. Co-dominance condition
    D. None of these

31. Which of the following laws of Mendel is well recognized across the world ?
    A. Law of segregation
    B. Law of independent assortment
    C. Both A & B
    D. None of above

32. A cross was made between two distinguished plants. for two characters, one allele was dominant over other gene for each character, then phenotypic ratio in F_2 generation would be according to:
    A. Law of independent assortment
    B. Law of dominance
    C. Law of gamete purity
    D. All of these

33. In which country, Mendel was born:
    A. USA
    B. India
    C. Austria
    D. Japan

34. Mendel studied which of the following subjects at University of Vienna ?
    A. Mathematics
    B. Philosophy
    C. Physics
    D. All of these

35. Mendel's Research paper title was one of the following:
    A. Experiment in Plant hybridization
    B. Experiment in pea plant
    C. Experiment of sorghum
    D. None of these

36. How many contrasting characters were studied by G. J. Mendel in *Pisum sativum* L.?

A. Fourteen B. Seven

C. Both A & B D. None of these

37. Phenotypic test cross ratio, which was Mendel for monohybrid cross?

A. 1:1 B. 1:1:1:2

C. 1:1:1 D. None of these

38. Phenotypic test cross ratio, which was observed by Mendel for dihybrid cross?

A. 2:2:2:2 B. 1:2:2:1

C. 1:1:1:1 D. All of these

39. Which of the following scientist coined the term "Genetics"

A. Bateson in 1905 B. Corren's in 1905

C. Muller in 1905 D. Ts Charmak in 1905

40. The term genetics was derived from which of the following language?

A. Greek B. Latin

C. Hindi D. Japanese

41. The term genetics means:

A. To grow B. To generate

C. To damage D. Both A & B

42. Which of the following branch comes in biological science:

A. Zoology B. Botany

C. Genetics D. All of these

43. Who rediscovered Mendel's laws in 1900?

A. Correns C B. Hugo de Vries

C. Ts Charmak D. All of these

44. Mendel paper was published in which of the following languages?

A. German B. English

C. Chinese D. None of these

45. Who translated Mendel papers from German language into English?

A. Bateson B. Corren's

C. Both A & B D. None of these

46. Who is the founder of the journal of genetics?

A. Punnett and de Vries B. Muller and de Vries

C. Punnett and Bateson D. N. I. Vavilov and M.S. Swaminathan

47. Suppose, a gene has two allele (RR) and also another gene has two allele (rr); cross made between them and obtained F_1 show Rr in which R allele suppressed another allele. These condition is known as:

A. Dominance B. Co-dominance

C. Incomplete dominance D. All of these

48. Who gave the term character?

A. Bateson
B. Mendel
C. Muller
D. Katayam
E. None of these

49. Which of the following term was proposed by Gregor Johann Mendel?

A. Factor
B. Dominant
C. Recessive
D. All of these

50. Hugo de Vries was native of:

A. Germany
B. Holland
C. Austria
D. None of these

51. C. Correns was born at :

A. Germany
B. Holland
C. Austria
D. None of these

52. Eric Von Tschermack was native of:

A. Austria
B. Germany
C. Holland
D. None of these

53. Which symbol is used to represent a male gamete or male parent?

A. ♂
B. ♀
C. Both A and B
D. None of these

54. Botanical name of four 'O' clock plant is:

A. *Mirabilis jalapa*
B. *Drosophila melenogester*
C. *Ratus ratus*
D. All of these

55. Which of the following is an example of co-dominance?

A. Blood group in human
B. Flower color in four 'O' clock plant
C. Both A & B
D. None of these

56. Which of the following tests was used by Mendel for testing of phenotypes of plants in his experiments?

A. Back cross
B. Test cross
C. Both A & B
D. None of these

57. Suppose, a cross made between smooth seeds having plant (SS) and another wrinkled seeds having plant (ss), results F_1 plant (Ss) obtained. After these results, a scientist take F_1 plant and cross made between F_1 and their wrinkled types plant. According above cross (F_1 × wrinkled types plant), which of the following ratio is produced:

A. 1:1
B. 3:1
C. 9:3:3:1
D. None the these

58. Back cross refers to:

A. Cross between F_1 and their either parent B. Cross between $F_1 \times F_1$

C. Both A & B D. None of these

59. A gene, which is responsible for the expression of two or more characters is known as ?

A. Pleiotropic gene B. Pleiotrophism

C. Dominant gene D. None of these

60. A condition, in which a single gene controls the expression of various characters in an individual. This types condition is known as:

A. Deletion B. Pleiotrophism

C. Dominant gene D. None of these

61. Which of the following, genetic constitution represents double recessive condition:

A. aabb B. Aaaa

C. AaBb D. AaBB

62. In diploid organism, homozygote is represented by:

A. Double symbol B. Single symbol

C. Both A & B D. None of these

63. Who proposed the theory of chromosomal inheritance in 1903?

A. Sutton and Bovari B. Sutton and Morgan

C. Boveri and Morgan D. None of these

64. Which of the following hypothesis has relationship between chromosome and Mendelian factor?

A. Laws of Mendelian B. Theory of mutation

C. Theory of chromosomal inheritance D. None of these

Terms

- **Crossing:** A synonym of mating means a process by which pollen reached on stigma by artificially and the product of crossing is known as hybrid.
- **Direct cross:** A cross made between two individuals or one way cross.
- **Genotypes:** Genetic makeup of an individual.
- **Phenotypes:** Visible expression of an individual.
- **Genetic variation:** The variation due to differences in genetic constitution of plants or organisms.
- **Homozygotes:** Individual having identical alleles of a gene or more genes.

- **Heterozygotes:** Individual having dissimilar alleles of a gene or more genes.
- **Pleiotropic gene:** A gene is responsible for the expression of two or more characters and condition is known as Pleiotropism.
- **Dominance:** A condition arises when one trait appears and other does not.
- **Monohybrid cross:** Cross made between parents differing in only one traits and its product is known as monohybrid.
- **Gamete:** A reproductive cell or sex cell.
- **Recessive trait:** A trait is expression masked by another.
- **Dominant trait:** A trait expresses itself in heterozygous condition.
- **Co-dominance:** A condition arises by which both alleles of a pair are fully expressed in a heterozygote.
- **Semi-dominance:** Absence of complete dominance.
- **Lethal gene:** Genes may affect viability as well as the visible traits of an organism. In other hands, gene responsible for causing the death of an organism.
- **Forward genetics or classical genetics:** Study of genes or DNA which is responsible to what is kind of phenotypic expression of an individual.
- **Reverse genetics:** Study of phenotypic expression which is caused by genes.
- **Allele:** Alternative forms of a gene.
- **Backcross:** Cross of a F_1 with either of the two parents.
- **Segregation:** Separation of alleles from each other or separation of parental and maternal chromosomes from each other at meiosis.
- **Amorph:** A mutation that obliterates gene function.
- **Atavism:** Reappearance of ancestral characters after several generations.
- **Epigenetics:** It refers to the expression of non genetic causes of a phenotype.
- **Hemizygous:** Condition arises by which only one allele of a pair is present resulting by heterozygous deletion.
- **LD:** Lethal dose causes lethality in population.
- **Mendelian ratio:** A ratio of progeny phenotypes imitating the operation of Mendel's law.

ANSWERS

Chapter 4: Law of inheritance

1. A	16. A	31. C	46. C	61. A
2. B	17. B	32. A	47. A	62. A
3. A	18. A	33. C	48. A	63. A
4. D	19. A	34. D	49. D	64. C
5. B	20. B	35. A	50. B	
6. A	21. A	36. B	51. B	
7. A	22. D	37. A	52. A	
8. D	23. A	38. C	53. A	
9. A	24. A	39. A	54. A	
10. A	25. A	40. A	55. A	
11. D	26. A	41. D	56. C	
12. A	27. B	42. D	57. A	
13. C	28. A	43. D	58. A	
14. A	29. B	44. A	59. A	
15. A	30. C	45. A	60. B	

Chapter 5

Epistasis

1. Epistasis is also known as gene interaction may caused by :

   A. Presence of homozygous recessives/dominant allele at one gene pair

   B. Only presence of a dominant allele at one gene pair

   C. Only presence of homozygous recessives allele at one gene pair

   D. None

2. Suppose, a cross made between two species of beans, one having white flower and the other having purple flowers to obtain all purple in F_1 and F_2 produced white flowers to purple flowers. Which condition/effect arise due to :

   A. Additive — B. Dominance

   C. Epistasis — D. All of these

3. A wide adaptable variety of characters influenced by additive relations among more gene pairs; character called :

   A. Qualitative character — B. Major character

   C. Both A & B — D. Quantitative character

4. Synergistic epistasis condition arise due to :

   A. When a double mutation is more effective than other single mutation

   B. Single mutation is more deleterious then other double mutations

   C. Additive effects of the single mutants

   D. All of these

5. Gene interaction can be clarified as :

   A. Allelic gene interaction — B. Non-allelic gene interaction

   C. Inter-allelic — D. Both A & B

6. Genes affect each other's expression known as :

   A. Gene interaction — B. Allelic gene interaction

   C. Non-allelic gene interaction — D. All of these

7. Which ratio arise in F_2 of test cross when supplementary gene interaction occurs :

A. 1:1:2 B. 1:1:1

C. 3:1 D. 1:3

8. Ratio arise in F_2 of test cross when complementary gene interaction occurs :

A. 1:3 B. 1:2

C. 1:1:3 D. None

9. In case of gene interaction of inhibitory, 13:3 ratio is obtained in F_2; after testcross of F_2 the ration of genotypes will be produced :

A. 1:3 B. 1:4

C. 1:2:1 D. None

10. The ratio of testcross in relation of duplicate factor will be observed :

A. 3:1 B. 1:3

C. 2:1 D. None

11. Which of the following condition arise a new character in organism ?

A. Incomplete dominance B. Complete dominance

C. Co-dominance and over dominance D. All of these

12. Lethality arise due to :

A. Lethal gene B. Lethal dose

C. Both A & B D. None

13. In maize, albinism is example of :

A. Lethal factor B. Gene interaction

C. Both A & B D. None

14. In case of mice, who discovered lethal gene expression :

A. Cuenot (1905) B. Cuenot (1811)

C. Muller (1910) D. None

15. Examples of Modifying genes :

A. White spotting on coat of pig B. Coat color in mice

C. Both A & B D. None

16. Gene which affects or alters the expression of some other non-allelic gene is known as

A. Modifying Genes B. DNA

C. Major genes D. Protein

17. Example of multiple alleles :

A. Coat color in mice.

B. Coat color in rabbit and Rh factor

C. Eye color in Drosophila and blood group (ABO) in human.

D. All of these

18. Multiple phenotypic effect of a single gene is called

A. Co-dominance B. Dominance

C. Pleiotropism D. None

19. How many genes are responsible to determine the duplicate gene interaction ?

A. Two dominant gene B. Two recessive gene

C. One dominant and one recessive genes D. None

20. Which example represent the duplicate gene interaction (15:1) in rice

A. Non-floating habitat of rice B. Awn character in rice

C. Both A & B D. None

21. The ratio of complementary gene action is

A. 8:7 B. 9:7

C. 15:1 D. 9:3:3:1

22. Complementary gene action is also known as :

A. Duplicate recessive epistasis B. Recessive epistasis

C. Dominant epistasis D. None

23. Example of duplicate recessive epistasis :

A. Flower color in sweet pea B. Awn traits in rice

C. Anthocyanin pigment in rice D. None

24. Which of the following ratio show inhibitory gene action ?

A. 13:3 B. 9:3:3:1

C. Both A & B D. None

25. Best example of inhibitory gene interaction is

A. Aleurone color in maize B. Flower color in sweet pea

C. Both A & B D. None

26. In case of polymeric gene interaction, ratio of genotypes will be obtained in F_2 generation:

A. 9:6:1 B. 9:7

C. 9:3:6 D. None

27. In barley, length of awns is affected by which genes :

A. Recessive genes B. Dominant genes

C. Co-dominant genes D. All of these

28. If the awns length of barley has been governed by dominant genes but contrasts genes are pairs each other's results :

A. Long awns barley B. Awns less barley

C. Middle length of awns in barley D. None

29. Fruits shape of summer squash is a most important example of :

A. Polymeric gene interaction B. Inhibitory gene action

C. Both A & B D. None

30. Which types of gene interaction is responsible for the governess of aleurone color in maize ?
A. Complementary gene action
B. Duplicate dominant gene action
C. Both A & B
D. Supplementary gene action

31. Masking gene action is found in :
A. Color of seed in jowar
B. Fruit color in summer squash
C. Both A & B
D. None

32. How many genes involved to governing pigment glands on cotton ?
A. Two
B. Three
C. Four
D. Five

33. What results will be obtained when crossed between glandular cotton plant with glandless plant ?
A. Glandular plant
B. Glandless plant
C. Both A & B types plants
D. None

34. A group of genes which altered the phenotypic expression /effect create by few genes called :
A. Modifying genes
B. Minor gene
C. Both A & B
D. Enhancer genes

35. How many genes responsible to production of HCN in clovers ?
A. 2
B. 3
C. 4
D. 5

36. Pleiotropic effect are found in ?
A. Maize
B. Wheat
C. Rice
D. All of these

37. In case of wheat, a gene governed awns in Onas (variety of wheat) results increase the yield and grain weight. This genes show which type of interaction ?
A. Pleiotropic interaction
B. Duplicate gene interaction
C. Recessive gene interaction
D. All of these

38. Pleiotropic effect is found in :
A. All plant species
B. Human beings
C. Both A & B
D. None

39. Hypostatic genes are those genes suppress by :
A. Hyperstatic genes
B. Epistatic gene
C. Both A & B
D. None

40. Epistatic gene is :
A. Express on other gene
B. Suppress by other genes
C. Do not affect any others genes
D. None

41. Epistatic gene interaction is also influenced by :

A. Penetrance
B. Expressivity
C. Environmental conditions
D. All of these

42. Uniformity expressivity is exhibited in :

A. Qualitative traits
B. Quantitative traits
C. Bio-chemical traits
D. All of them

43. The tem epistasis was coined by :

A. Bateson, 1909
B. Bateson, 1900
C. Bateson, 1899
D. None

44. One gene express and another gene suppress condition arise in an organ due to :

A. Inter-gene interaction
B. Intra-allelic gene interaction
C. Pleiotropic expression
D. All of these

45. Who observed gene interaction in comb shape in poultry ?

A. Bateson and Punnett
B. Bateon and Muller
C. Muller and Mc Clintock
D. None

46. Epistasis differ to dominance base on :

A. Two or more loci involve
B. Recessive also exhibited masking effect
C. Both A & B
D. None

47. Epistasis is similar to dominance based on :

A. Nuclear gene
B. Gene effect same traits
C. Both A & B
D. None

48. Pleiotropism is a condition arise due to :

A. Only one gene express more than one traits
B. Only recessive gene express this condition
C. Both A & B
D. None

49. Antagonistic pleiotropy hypothesis was developed by :

A. G. C. Williams, 1957
B. Bateson, 1909
C. Both A & B
D. None

50. Well-known examples of pleiotropy include :

A. Albinism
B. Sickle-cell anemia
C. Both A & B
D. Height of human

51. The term "pleiotropie" was coined in a 1910 :

A. Festschrift
B. Bateson
C. Heacle
D. None

52. Which language is closely related to pleiotropy ?

A. Greek
B. Latin
C. Russian
D. None

53. The degree of phenotypic expression is called expressivity affected by :

A. Temperature
B. Humidity
C. Both A & B
D. None

54. In mammals, *C* gene is associated to the :

A. Production of tyrosinase
B. Albino
C. Both A & B
D. None

55. According to Silvers, how many gene affect coat color in mice :

A. More than 60
B. 40
C. 17
D. None

56. Which genetic material studied by Dobzansky in 1927 and revealed that the white eye of this organism may affect the shape of sperm storage organ in female as well as their structure ?

A. Mice
B. Moth
C. Drosophilla
D. None

57. Who reported lethal gene in *Snapdragon (Antirrhinum majus)* during 1907 ?

A. Cuenot
B. E. Baur
C. Bateson
D. Punneett

58. Bombay phenotype in ABO blood groups is a good example of :

A. Recessive epistasis
B. Duplicate
C. Inhibitory
D. None

59. Who described Preferential segregation in maize ?

A. M Rhoades, 1942
B. Mc Clintock, 1983
C. Morgan, 1910
D. None

60. Preferential segregation in maize cause due to :

A. Abnormalities in chromosome no 10 in maize
B. Extra block of heterochromatin near the end of its long arm forming a knob
C. Both A & B
D. None

Terms

- **Complementary recessive epistasis:** A condition arises by pairs of recessive alleles at same or different loci masks the expression of dominant alleles at two loci. This is known as complementary gene action.
- **Epistatic gene:** A gene that overcomes the expression of other gene. In other words, interaction between non allelic genes in which on gene suppresses the expression of other gene, called epistasis and gene is known as epistatic gene. These types of gene shows both penetrane and expressivity effect of genes.
- **Expressivity:** The degree of visibility or phenotypic expression of a penetrant gene.
- **Gene interaction:** A phenomenon that consists of the effect of one gene being dependent on the presence of one or more 'modifier genes' (genetic background).
- **Hypostatic gene:** A gene expression is suppressed by another gene. In other words, hypostatic gene is that one whose expression is masked by another.
- **Incomplete penetrance:** When specific gene does not manifest its effect in all individuals which carry this gene, it is known as incomplete penetrance.
- **Inhibitory epistasis:** A condition arises by dominant allele at one locus when they produced their effects on both alleles at second locus.
- **Masking gene action:** A condition arises by dominant allele when they mask both alleles at other loci. It is also known as simple epistasis.
- **Penetrance:** The frequency of a gene which produces a visible expression in an individual.
- **Quantitative character:** Trait defined as a wide adaptable variety of characters influenced by additive relations among more gene pairs.
- **Supplementary gene action or recessive epistasis:** A condition arises by recessive allele at one locus when they mask the expression of both alleles at other loci.
- **Uniform expressivity:** The degree of uniform expression of a gene in all population. This type expressivity exhibited by quantitative gene.

ANSWERS

Chapter 5: Epistasis

1. A	16. A	31. C	46. C
2. A	17. D	32. A	47. C
3. D	18. C	33. A	48. A
4. A	19. A	34. C	49. A
5. D	20. C	35. A	50. C
6. D	21. B	36. D	51. A
7. A	22. A	37. A	52. A
8. A	23. A	38. A	53. C
9. A	24. A	39. B	54. A
10. A	25. A	40. A	55. A
11. D	26. A	41. D	56. C
12. C	27. B	42. A	57. B
13. C	28. B	43. A	58. A
14. A	29. A	44. D	59. A
15. C	30. A	45. A	60. C

Chapter **6**

Maternal Effects and Cytoplasmic Inheritance

1. Cytoplamic inheritance can be determined through :

   A. Direct crosses B. Reciprocal crosses

   C. One way crosses D. All of these

2. Cytoplasmic inheritance is controlled by :

   A. Non-nuclear genome B. Plasmagene

   C. Cytoplamic gene D. All of these

3. Which gamete contributes all the cytoplasm to the zygote ?

   A. Female gamete B. Male gamete

   C. Both A & B D. None

4. Chromosomal gene in a particular cytoplasm could be incorporated through :

   A. Repeated back crosses B. Test crosses

   C. Direct crosses D. None

5. Which of the following produces direct effect on zygote development?

   A. Maternal cytoplasm B. Nuclear genome

   C. Both A & B D. DNA

6. Non-Medelian segregation pattern is observed in :

   A. Extra chromosomal inheritance

   B. Maternal inheritance

   C. Cytoplasmic inheritance and uniparental inheritance

   D. All of these

7. Maternal effect is observed in :

   A. Water flea B. Snail

   C. Flour moth D. All of these

8. Nuclear gene of mother determines the phenotype of the progeny :
   A. Maternal effect
   B. Male gamete effect
   C. Both A & B
   D. Nuclear chromosomal effect
9. True cytoplasmic inheritance is due to independent self replicating nucleic acid present in :
   A. Mitochondria
   B. Chloroplast
   C. Leucoplast in human
   D. All of these
10. Extra chromosomal factor is :
    A. Plasma genes
    B. Plasmids
    C. Cytogenes
    D. All of these
11. Who reported the inheritance patterns controlled by cytoplasmic genes or plasma genes in 4'O clock in 1909 ?
    A. Correns
    B. Ephrussi
    C. Sorieul
    D. R Sanger
12. Antibiotic resistance in reciprocal crosses is controlled by :
    A. Non chromosomal genes
    B. Nuclear genome
    C. Both A & B
    D. Mendelian gene
13. Neutral petite is controlled by :
    A. An extra chromosomal gene
    B. Segregational petites
    C. Both A & B
    D. None
14. The frequency of petite mutation is :
    A. 10^{-5} per generation
    B. 10^{-3} per generation
    C. 10^{-8} per generation
    D. None
15. Suppressive petite is caused by :
    A. Large deletion in mitochondria
    B. Large deletion of nuclear genome
    C. Both A & B
    D. Addition of mitochondrial gènome
16. Plasmagene is a part of DNA codes protein for a phenotypic expression and does not follow which on of the following inheritance pattern ?
    A. Conventional Mendelian pattern
    B. Extra-chromosomal pattern
    C. Both A & B
    D. None
17. Plant mitochondrial genomes have only following subunit of ATPase ?
    A. atpA
    B. atp6 and atp0
    C. Both A & B
    D. atPE
18. Egg cytoplasm is responsible for :
    A. Extra chromosomal inheritance
    B. Nuclear gene inheritance
    C. Both A & B
    D. Mendelian inheritance

19. Who observed the pattern of CMS inheritance in maize in 1933 ?
A. Rhoades B. Honjo
C. Ossava D. Kanji *at al.*,

20. Restorer nuclear gene is able to suppress :
A. Male sterility effect of the cytoplasm B. To produce the Mendelian inheritance
C. Found in male sterile line D. All of these

21. Which type of male sterility system is operative in maize ?
A. CMS-S B. CMS-T
C. CMS-C D. All of these

22. Paramicin is :
A. A toxic substance
B. Water soluble and diffusible
C. Production depends upon the presence of cytoplasmic bacteria
D. All of these

23. When the cross is made between organisms carrying traits caused by mtDNA, then its transmission will be according to :
A. Mendelian inheritance pattern B. Nuclear genome inheritance pattern
C. Both A and B D. Cortical inheritance patterns

24. Who discovered mutant strain of petites in baker's yeast ?
A. Ephrussi *at al.*, B. Corren's
C. Muller D. Jensen

25. Who demonstrated the inheritance of two alternative traits in strains of *Paramecium Aurelia* in 1943 ?
A. Sonneborn B. Ephrussi
C. Frederick D. None

26. Who discovered the pigmentation in flour moth (*Ephestia kuhniella*) in 1936 ?
A. Sonneborn B. Caspari
C. Ephrussi D. Corren's

27. Essential cell organelles of cytoplasmic inheritance are :
A. Mitochondria B. Cytoplasm
C. Both A & B D. Nucleus

28. Who proposed the theory of plasma gene ?
A. Sonneborn B. Ephrussi
C. Frederick D. None

29. Who proposed the circular map and several gene pair localized on chromosome in 1972 ?
A. Ruth Sager B. Corren's
C. Teisser D. None

30. Who discovered a true breeding strain of Drosophilla sensitive to CO_2 in 1958 ?
    A. L'Heritiei and Teisser
    B. Ruth Sager
    C. Sonneborn
    D. Ephrussi
31. Which of the following characteristics is controlled by mitochondrial genome ?
    A. Cytoplasmic male sterility
    B. Poky in *Neurospora*
    C. Petite in yeast
    D. All of these
32. Barstar is encoded by which of the following ?
    A. RNase inhibitor
    B. RNase enhancer
    C. Both A & B
    D. None
33. Both *barnase* and *barstar* genes are isolated from :
    A. Bacteria
    B. Fungi
    C. Algae
    D. Fern
34. Both *barnase* and *barstar* genes were isolated from :
    A. *Bacillus amyloliquefaciens*
    B. *Chlamydomonas*
    C. *Paramecium*
    D. All of these
35. Which of the following gene is responsible for creation of male sterility in *Brassica spp.?*
    A. Barnase
    B. Barstar
    C. Both A & B
    D. None
36. The pattern of inheritance of cpDNA and mtDNA are influenced by :
    A. Vegetative sorting
    B. Differences in the rate of replication of organelle DNAs
    C. Both A & B
    D. None
37. Segment of DNA migrated from one organelle to other is known as :
    A. Mobile element
    B. Transposon
    C. Promiscuous DNA
    D. All of these
38. Which of the following feature is not found in cytoplasmic organelle genome?
    A. Observed circular DNA in organelle genome
    B. Observed linear DNA in organelle genome
    C. Encoded all RNA species
    D. Occurrence the transcription and translation
39. *Iojap* is a type of variation controlled by :
    A. A recessive nuclear genome
    B. A dominant nuclear genome
    C. Both A & B
    D. Psedo-gene

40. Biparental plastids transmission is found in :

A. *Pelargonium* B. *Oenothera*

C. Medicago D. All of these

41. Petite mutation of *S. cerevisiae* is :

A. Nuclear B. Neutral

C. Suppressive D. All of these

42. Iojap trait in corn is controlled by a gene located on :

A. Chromosome no. 7 B. Chromosome 8

C. Chromosome 1 D. None

43. Who mapped a gene controlling the Iojap trait in corn and presence on chromosome no 7 in 1946?

A. M. Rhoades B. Baur

C. Correns D. None

44. The mating between poky female and normal male of *Neurospora* resulting the production of:

A. All poky *Neurospora* B. All normal *Neurospora*

C. 50 % pokyness *Neurospora* D. None

45. Who reported pattern of plastid inheritance in *Oenothera* ?

A. Renner B. M. Rhoades

C. Both A & B D. Baur

46. Who described the iojap trait in maize?

A. Jenkin B. M. Rhoades

C. Sager D. Palmer

47. Which of the following, governs the leaf variation in four 'O' clock plants?

A. Nuclear gene B. Plasmagene

C. Environmental gene D. All of these

48. Which of the following cytoplasm is linked with southern corn leaf blight?

A. CMS-T B. CMS-C

C. Both A & B D. CMS-S

49. Which of the organelle of plant or cell contains iojap gene in maize :

A. Plastid B. Mitochondria

C. Ribosome D. E. R.

Terms

- **Maternal effect:** The visible expression of progeny is being affected by their mother parent. In other words, the effect on progeny is determined from nuclear genes arises their female parent.
- **Cytoplasmic inheritance:** The transmission of traits from parent to their offspring can be determined by plasmagenes or extra-nuclear gene or cytoplasmic gene.
- **Non Mandelian segregation:** The separation of allele represented the extra nuclear inheritance or maternal inheritance or uniparental inheritance.
- **Cortical inheritance patterns:** A pattern of inheritance in which the cross can be made between organism carrying traits caused by mtDNA transmitted.
- **Variegation:** The presence of white or yellow spots in variable size on green leaves.
- **Plasmon:** The sum total of gene found in cytosols is called Plasmon.
- **Plastid inheritance pattern:** The inheritance governed by gene present in plastid.
- **Mitochondria inheritance:** The transmission of traits from parents to their offspring determined by mtDNA.
- **Pearson marrow-pancreas syndrome:** A syndrome governed by a mutation in mtDNA.
- **Mitochondrial DNA:** Mostly circular DNA found in mitochondria represented self replication.
- **Paramecin:** A toxin substance produced by *Paramecium*.
- **Barnase gene:** A gene is responsible to creation of male sterility in *Brassica spp.*
- **Cytochrome:** A type of protein found in mitochondrial membrane and play key role in oxidative phosphorylation of ADP to from ATP.
- **Cytosol:** A fluid portion of cytoplasm excluding organelles.
- **Matroclinous inheritance:** A pattern of inheritance in which all offspring have phenotype as mother parents.

ANSWERS

Chapter 6: Maternal effects and cytoplasmic inheritance

1. B	16. A	31. D	46. A
2. D	17. C	32. A	47. D
3. C	18. A	33. A	48. A
4. A	19. A	34. A	49. A
5. A	20. A	35. A	
6. D	21. D	36. C	
7. D	22. D	37. D	
8. A	23. D	38. B	
9. D	24. A	39. A	
10. D	25. A	40. D	
11. A	26. B	41. D	
12. A	27. C	42. A	
13. A	28. A	43. A	
14. A	29. A	44. A	
15. A	30. A	45. A	

Chapter **7**

Population Genetics

1. The frequency of a gene and genotypes is dependent on which of the following factor:
   A. Environmental factor B. Gene introgression
   C. Mutation D. Both B & C
2. Which of the following factor influences the frequencies of gene and genotypes in a population generation after generation ?
   A. Mutation B. Migration
   C. Random drift D. All of these
3. Evolutionary factor is:
   A. Mutation and migration B. Selection and random drift
   C. Both A & B D. Linkage
4. Which of the following study is associated with genetic architecture of a population ?
   A. Population genetics B. Evolutionary genetics
   C. Both A & B D. None
5. The study of genes within the populations is known as
   A. Evolutionary genetics B. Cytology
   C. Genomics D. Forward genetics
6. Study of the frequencies of genotypes within the panmictic population is :
   A. Evolutionary genetics B. Population genetics
   C. Both A & B D. Genomics
7. The sum total of genes in reproductive gametes of a Mendelian population is known as:
   A. Gene pool B. Genotypes
   C. Gene frequency D. None
8. Who proposed the term Mendelian population in 1935 ?
   A. Sewall Wright B. Weinberg
   C. Dobzhansky D. Hardy, G. H.

9. Who observed that the genes and genotypes frequencies remain constant in random mating population generation after generation in 1908, if any evolutionary factor is not present there ?

A. G. H. Hardy and W. Weinberg B. Sewall Wright

C. Ephrussi D. None

10. Who proposed the genetic equilibrium of population?

A. G. H. Hardy and W. Weinberg B. Sewall Wright

C. Ephrussi D. None

11. The term meiotic drive was coined by:

A. Flemming B. Strasburger

C. Sandler and Novitski D. None

12. Which of the following law states that gene and genotypes remain constant generation after generation?

A. Mendelian law B. Hardy-Weinberg law

C. Law of chromosomal inheritance D. Linkage disequilibrium

13. Which of the following is Hardy–Weinberg equation?

A. P^2 (AA) + 2pq (Ab) + q^2 is equal to one

B. P^2 (AA) + 2pq (Ab) + q^2 is equaled to Zero

C. Both A & B

D. P^2 (AA) + 2pq (Ab) + q^2 is always less than one

14. Which of the following assumption is not based on population genetics?

A. Population size is infinite

B. Population arise through random mating

C. Male and female gametes have dissimilar allele frequency

D. There is no mutation, migration, selection and drift

15. Which factor does not affect panmictic population?

A. Mutation B. Migration

C. Selection D. Linkage

16. Fitness is known as:

A. Adaptive value B. Selective value

C. Both A & B D. None

17. Capacity of genotype to produce more offspring as compared to another under same conditions is called:

A. Adaptive value B. Selective value

C. Fitness D. All of these

18. Which term is related to reproductive value of the genotype?
    A. Fitness B. Adaptive value
    C. Selective value D. All of these
19. The adaptive value is designed by which of the following :
    A. A B. w
    C. S D. None
20. Range of fitness or selection coefficient varies between :
    A. 0 to 1 B. More than one
    C. Only zero D. None
21. The fitness of genotypes is equal to Zero. Which of the following condition is assumed for the offspring of individual:
    A. Not survival population B. More lethal
    C. Both A & B D. None
22. Which of the following symbol is designed for selection coefficient?
    A. S B. W
    C. rd D. h
23. The term Natural selection was proposed by :
    A. Darwin B. Johansson
    C. Mather D. None
24. The concept of fitness is :
    A. Central to natural selection B. Central to selective selection
    C. Central to artificial selection D. All of these
25. Disruptive selection can be a precursor to :
    A. Speciation B. Positive selection
    C. Stabilizing selection D. All of these
26. Purifying selection acts to :
    A. Remove the genetic variation from the population
    B. Attrition heterozygosity
    C. Both A & B
    D. Creation of breeding line
27. The gene frequency gets changed due to :
    A. Random mating B. Selection
    C. Out-breeding D. Both B & C
28. The genotype frequency is changed due to :
    A. Uniting male or female gamete has equal chance for each
    B. Migration of genotype from one population to other
    C. Selection coefficient is equal to zero
    D. None

29. Panmictic population developed by which of the following procedure:
    A. Random mating
    B. Genetic assortative mating
    C. Phenotypic assortative mating
    D. Phenotypic disassortative mating
30. Panmictic population is known as:
    A. Mendelian population
    B. Natural population
    C. Outbreed
    D. None
31. Which of the following factor is nondirectional factor affecting the gene or genotype frequencies ?
    A. Genetic drift
    B. Mutation
    C. Selection
    D. Migration
32. Which of the following factor is directional factor affecting the gene or genotype frequencies of population ?
    A. Mutation
    B. Selection
    C. Migration
    D. All of these
33. Who proposed the use of a population of finite size in population genetics ?
    A. Hardy and Weinberg
    B. Wright and Fisher
    C. Johansson
    D. None
34. Which of the following can be used to study frequencies of each allele of pair of genes as well as the frequencies of homozygote and heterozygotes in the population ?
    A. Population genetics
    B. Forward genetics
    C. Linkage disequilibrium
    D. None
35. Which of the following steps is involved to deduction of the Hardy-Weinberg law ?
    A. From parent to progeny, gametes they produce
    B. From union of gametes to genotypes in zygote they produce
    C. From the genotypes of the zygote to gene frequencies in the progeny generation
    D. All of these
36. Which factor affects the gene frequency in the population ?
    A. Sampling error
    B. Random mating
    C. Both A & B
    D. None
37. The changes in the gene frequency due to selection depends on which of the following :
    A. Degree of dominance with respect to the fitness
    B. Degree of dominance with respect to the adaptive value
    C. Degree of dominance with respect to the selective value
    D. All of these
38. The frequency of mutated gene in the population is :
    A. High
    B. Low
    C. Predefine
    D. None

39. Which is true in respect of genetic drift ?
 A. Magnitude the drift that occurs in each generation is independent
 B. Drift in each generation is random event
 C. The direction of drift cannot be predictable
 D. All of these

40. Fitness is measured by :
 A. The proportion contribution of the offspring to the next generation
 B. Gene value in the organism
 C. Both A & B
 D. None

41. Fitness indicates :
 A. Adoptive value B. Selective value
 C. Reproductive value D. All of these

42. The genotypic frequencies in the population follows :
 A. Binomial distribution B. Poisson approximation
 C. Beta distribution D. None

43. Wright-Fisher population model assume that :
 A. Fix population size and
 B. Synchronized and non overlapping generation
 C. Both A & B
 D. None

44. The Wright –Fisher population is affected by :
 A. Migration force B. Mutation force
 C. Genetic drift force D. None

45. Who used F-statistics in population genetics to describe the pattern of genetic divergence and variation in 1968 ?
 A. Sewall Wright B. R. A. Fisher
 C. Both A & B D. Darwin

46. Which of the procedure is used to measure genetic diverge among the individuals of a population ?
 A. D square statistic B. Study of co-ancestry or pedigree
 C. Both A & B D. None

47. Which of the following method is used to estimate of genetic distance ?
 A. Nei's method B. Cavalla-Sforza method
 C. Reyond *et al.*, method D. All of these

48. Which type of selection favors for accumulation single allele at a specific locus in a population ?
 A. Directional selection B. Balancing selection
 C. Both A & B D. None

49. A large no. of event distributed in the population is an indication of diversifying condition based on which of the following selection:
    A. Balancing selection B. Directional selection
    C. Both A & B D. None
50. The theory of coalescent assumes which of the following:
    A. No recombination B. No natural selection
    C. No gene flow D. All of these
51. Who developed the mathematical theory of the coalescent?
    A. John Kingman B. Sewall Wright
    C. R. A. Fisher D. None
52. Coalescent theory is a natural extension of :
    A. Classical population genetics B. Wright-Fisher model
    C. Both A & B D. Forward genetics
53. Which of the following responsible for gene flow between populations?
    A. Migration B. Selection
    C. Random mating D. Recurrent mutation
54. In which of the following, the smaller the population is more susceptible to change as random changes:
    A. Genetic drift B. Mutation
    C. Gene flow D. None
55. Hardy-Weinberg laws operates in:
    A. Synthetics B. Composites
    C. Both A & B D. None
56. Which factor is responsible for direction of evolutionary process ?
    A. Natural selection B. Genetic recombination
    C. Gene mutation D. Chromosomal aberration
57. The book "Processes of Organic Evaluation" written by whom of the following:
    A. Stebbins B. Sewall Wright
    C. JBS Haldane D. Ernst Mayer
58. Which of the following factor is a minor or major contributor to speciation ?
    A. Genetic drift B. Migration
    C. Mutation D. Directional selection
59. The natural theory of molecular evolution was introduced by:
    A. Motoo Kimura B. McCarthy
    C. Richard Lewontin D. Tomoko Ohta

60. Who proposed the shifting balance theory of evolution and concept of adaptive landscape in 1932?
A. Sewall Wright
B. Ernst Mayr
C. C. J. Goodnight
D. None

61. Who got credit to develop biological concept of species?
A. Ernst Mayr
B. Sewall Wright
C. Richard Lewontin
D. None

62. Who proposed BDM model of the evolution?
A. Muller
B. Dobzhansky
C. Bateson
D. All of these

63. Which of the following model is associated with genetic view of speciation ?
A. BDM model
B. Founder-flush speciation model
C. Both A & B
D. None

64. Which of the following represents a random mating population ?
A. $p + q = 1$
B. $F = 0$
C. Both A & B
D. None

65. Which is not true in respect of inbreeding ?
A. The allele frequencies remain constant but genotype frequencies change
B. Occurs when mating between genetically related occurs
C. $F = 1$ indicate complete inbreeding
D. Decrease the prepotency value

66. Autozygote is a condition to represent:
A. The allele at a locus are identical by descent
B. The ful increment the value of prepotency
C. Homozygotes
D. All of these

67. Which of the following equation is represented by the frequency of dominant homozytoes individual if inbreeding depression is involved ?
A. AA: $pF + p^2 (1 - F)$
B. AA: $pF + p^2 (1 + F)$
C. AA: $pF + q^2$
D. None

68. Fitness is associated with :
A. Viability
B. Fertility
C. Reproducibility
D. All of these

69. Which theory supported the rate of evolution is equal to the rate of mutation ?
A. Neutral theory of evolution
B. Synthetic theory of evolution
C. BDM model of evolution
D. None

Terms

- **Artificial selection:** Selection made by human for different purposes.
- **Balancing selection:** The selection favors to heterozygotes, both of the alleles will remain in the population even though they are deleterious in homozygote state. Some time it is known as selection with overdominance.
- **Behavior genetics:** It is concerned with the effects of genotypes on behavior and with the role that genetic differences play in the determination of behavioral differences in a population.
- **Bottleneck:** A brief reduction in size in population lead by random drift.
- **Concerted evolution:** The overall process of maintaining genetic homogeneity in such multigene family is called concerted evolution.
- **Evolutionary force or factor:** The force that creates the variability in a species results evolution *i.e.*, migration, mutation, genetic drift and selection.
- **Evolutionary genetics:** The study of genes related to evolution of the population.
- **Fitness:** The capacity of a genotype/gene to reproduce more offspring as comparison to other genotypes in same condition. Fitness is also known as selective value or adaptive value. It is designated as **w.**
- **Gene pool:** The sum total of genes or gametes reproduced within a species. In other hands, the sum total of genes in reproductive gametes of a Mendelian population considered as gene pool.
- **Genetic random drift:** A non-directional factor of evolution arises due to sampling error. In other hands, during reproduction, allele frequencies can be changed by chance. This factor is affected the genes or genotypes frequency of population.
- **Inbreeding coefficient (*F*):** It is the probability of two alleles in an individual are identical by descent from a common ancestor. Inbreeding coefficient of individual depends upon the genetic relationship between its parents.
- **Inbreeding:** A process of repeated self pollination exhibit the general effect of a non random mating.
- **Linkage disequilibrium (*d*):** The difference between the products of the frequencies of the two repulsion phase gametes and the two coupling phase gametes.
- **Migration or gene flow:** Introgression of gene from one plant or population to another.
- **Molecular evolution:** A process of change. The process of evolution of an organism involves the insertion, depletion and substitution of nucleotides in the DNA.

- **Mutation:** Change arises in an individual show discrete, heritable and sudden properties.
- **Natural selection:** Selection which operates in by nature automatically.
- **Natural theory of molecular evolution:** The theory that states the rate of evolution is completely determined by the force of mutation. This theory some time is called non-Darwinian theory of evolution.
- **Natural theory of selection:** Theory states that the rate of evolution is equal to the rate of mutation.
- **Out-breeding:** The mating between two unrelated individuals.
- **Panmictic population:** Population arises due to random mating.
- **Population genetics:** The study of gene and genotypes frequencies which cannot changed generation after generation; if there is no evolutionary factor *i.e.*, mutation, selection, linkage disequilibrium, migration and random drift.
- **Random mating:** Randomly uniting the male and females with each others.
- **Selection coefficient:** Numerical value of selection lies between 0-1. Selection coefficient is symbolized as S.
- **The synthetic theory of evolution:** The theory explain the information about transformation of a species by natural selection and splitting of a species into reproductively isolated subgroups. After some time, this process is known as speciation.
- **Pure line:** Lines/individuals exhibit same genetic constitution.

ANSWERS

Chapter 7: Population Genetics

1. D	16. C	31. A	46. C	61. A
2. D	17. D	32. D	47. D	62. D
3. C	18. D	33. B	48. A	63. C
4. C	19. B	34. A	49. A	64. A
5. A	20. A	35. D	50. D	65. D
6. C	21. C	36. A	51. A	66. D
7. A	22. A	37. D	52. C	67. A
8. C	23. A	38. B	53. A	68. D
9. A	24. D	39. D	54. A	69. A
10. A	25. A	40. A	55. C	
11. C	26. C	41. D	56. A	
12. B	27. D	42. A	57. A	
13. A	28. B	43. C	58. A	
14. C	29. A	44. C	59. A	
15. D	30. A	45. A	60. A	

Chapter **8**

Quantitative Genetics

1. What is quantitative genetics?
   A. Study of inheritance of minor gene and its variation
   B. Study of developmental genetics
   C. Study of major genes and its variations
   D. Study of whole genome
2. A branch of biological science which deals phenotypes of the traits which vary continuously (such as height or mass) is known as:
   A. Quantitative genetics B. Forward genetics
   C. Reverse genetics D. Developmental genetics
3. Which of the following is recently addition useful to analysis of evolution of organism?
   A. QTLs B. Gene effect
   C. Random drift D. Sampling design
4. Quantitative traits express themselves with type of variation:
   A. Continuous B. Discontinuous
   C. Bidirectional D. None
5. The structure is similar between the two species because of independent adaptation to common environmental pressures, called:
   A. Homology B. Convergence
   C. Homogeneous D. None
6. Morphological features of an organism that can be measured is called:
   A. Quantitative feature B. Major feature
   C. Both A & B D. None
7. Morphological features such as leaf shape, flower color or pubescence is known as:
   A. Quantitative feature B. Major feature or quantitative feature
   C. Both A & B D. None

8. Quantitative traits have:
   A. Continuous distributions, not discrete classes
   B. Polygenic effect
   C. Environmental factors effects also
   D. All of these
9. The square root of variance is called:
   A. Coefficient of variation B. Variance
   C. Standard deviation D. Median
10. Genetic variation is heritable because it can be passed from :
   A. Parent to offspring B. Additive gene to dominance
   C. Both A & B D. None
11. Genetic variation is passed from parent to offspring but can't fixed :
   A. Dominance gene interaction B. Additive interaction
   C. Both A & B D. None
12. The sum of the genetic and environmental variance is known as:
   A. Mean sum of square B. Phenotypic variance
   C. Correction factor D. None
13. The additive and non-additive genetic variance (VA and VD) together are known as:
   A. Genotypic or genetic variance B. Phenotypic variance
   C. Environmental variance D. Homogeneity
14. The proportion of genotypic variance to the phenotypic variance is known as:
   A. Heritability narrow sense B. Heritability broad sense
   C. Selection deferential D. Genetic gain
15. The proportion of additive genetic variance to the phenotypic variation is known as:
   A. Heritability narrow sense B. Heritability broad sense
   C. Selection deferential D. Genetic gain
16. The proportion of variability that can be passed on from parent to offspring is known as:
   A. Heritability B. Genetic advance
   C. Genetic gain D. Breeding value
17. The heritability estimate is a very useful to predict the response to:
   A. Artificial selection B. Natural selection
   C. Both A & B D. None
18. The sum of average allelic effects at all loci influencing the trait of interest is called:
   A. Breeding value B. Heritability
   C. Correlation coefficient D. None

19. Which of the following is not simple measurable statistic of variability?

A. Mean
B. Variance
C. Standard deviation
D. Coefficient of variance

20. Which of the following method is used for assessment of variability in plant breeding?

A. Range, standard deviation and Coefficient of variation
B. Metroglyph analysis
C. D-square analysis
D. All of these

21. Use of various statistical principles or methods to study the biological problems is known as:

A. Biometrics
B. Biometrical genetics
C. Forward genetics
D. None

22. A branch of science which deals the study of minor gene or metric traits is called:

A. Quantitative genetics
B. Genomics
C. Proteomics
D. Biometrics

23. Who was the founder of quantitative genetics?

A. R. A. Fisher
B. Dalton
C. Francis Galton
D. Karl Pearson

24. Estimation of quantitative traits are based on the following:

A. Mean
B. Variance
C. Covariance
D. All of these

25. Population genetics differ quantitative genetics in respect of:

A. Population genetics deals genes and genotypes frequency
B. Population genetics doesn't measure polygenic inheritance
C. Population genetics deals also oligogenic variation
D. All of these

26. Who was the founder of population genetics?

A. Hardy-Weinberg
B. R. A. Fisher
C. Morgan
D. None

27. Who coined the term eugenics and proposed the idea of improving the physical and mental makeup of human species by selective mating of marriage partners?

A. F. Galton
B. Darwin
C. Smith
D. None

28. Who used regression coefficient for estimation of the relationship between the heights of father and their son?

A. Galton
B. Kearsey
C. Dudley
D. Dewey

29. Which is truly matched:

Column I	*Column II*
A. Galton	*i.* Concept of correlation
B. Smith	*ii.* Discriminant function analysis
C. Cavalli	*iii.* Restricted selection index
D. Kempthorne and Nordkog	*iv.* Joint scalling test

A. a-i, b-ii, c-iv, d-iii
B. a-ii, b-iii, c-iv, d-iii
C. a-iii, b-iv, c-i, d-ii
D. a-iv, b-ii, c-i, d-iii

30. Which of the following technology is used for the study of anthropology and psychometry?
A. D-square analysis
B. Metroglyph
C. Both A & B
D. None

31. Which of the following technique is used for the assessment of varietal adaptability?
A. Triple test cross
B. Stability analysis models
C. Quadriallel analysis
D. None

32. D square statistics was developed by Mahalanobis, P. C. in 1928 working on:
A. Chinese head measurement
B. Assessment of varietal assessment
C. Both A & B
D. Measures of dispersion

33. Selection is more efficient to :
A. low genetic advance with high heritability of traits
B. High genetic advance with high heritability of traits
C. Low variability of population
D. None

34. Which of the following techniques is used to assess the phenotypic variability in the large gene pool?
A. D-square analysis
B. Metroglyph analysis
C. Both A & B
D. None

35. Which of the following selection is more efficient to improve the yield of crops?
A. Direct selection
B. Indirect selection
C. Bidirectional selection
D. None

36. Which of the following biometrical techniques is used to evaluate several single crosses, simultaneously?
A. Diallel analysis
B. Partial diallel analysis
C. Line X tester analysis
D. All of these

37. Which of the following techniques is used for evaluation of several three way crosses, simultaneously?
A. Triallel analysis
B. Diallel analysis
C. Qudriallel analysis
D. All of these

38. Several double crosses can be evaluated by which of the following:
A. Quadriallel analysis
B. Line X tester analysis
C. Diallel analysis
D. None

39. A measures of the tendency of individual genotypes in a population to vary from one another:
A. Variability
B. Correlation
C. Mean
D. None

40. Main source of genetic variability is:
A. Natural mutation
B. Natural out-crossing
C. Recombination
D. All of these

41. Which technique represents the variability of the characters by glyph on the graph?
A. Metroglyph analysis
B. D-square analysis
C. Cluster analysis
D. None

42. Which word is associated with D-square analysis?
A. Cluster
B. Glyph
C. Both A & B
D. None

43. A group of most identical genotypes is called:
A. Cluster
B. Gene pool
C. Base material
D. Land race

44. The proportion of average of phenotypic variables is known as:
A. PCV
B. GCV
C. ECV
D. None

45. Which of the following approach is considered as semi-graphical approach?
A. Metroglyph analysis
B. D square analysis
C. Both A & B
D. Covariance

46. First order statistics is used for analysis of:
A. Combining ability effects
B. Heterosis
C. Path and correlation
D. Both A & B

47. Second order statistics is used for analysis of:
A. Discriminant function, correlation and path coefficient
B. Correlation, path and genetic advance
C. Combining ability variance and heritability
D. All of these

48. First degree statistics includes:
A. Means
B. Variance
C. Covariance
D. None

49. Second degree statistics includes:
A. Covariance
B. Variance
C. Both A & B
D. None

50. The measures of regression coefficient was developed by :
A. Fisher
B. Galton
C. Pearson
D. None

51. The index of genotypic covariance to phenotypic covariance is called :
A. Co-heritability
B. Heritability
C. Regression
D. None

52. The sum of genotypic covariance and environmental covariance is known as:
A. Phenotypic covariance
B. Co-heritability
C. Genetic advance
D. None

53. The D square analysis is based on:
A. Variance
B. Covariance
C. Both A & B
D. Mean

54. Which technique is based on first order statistics?
A. D-square analysis
B. Metroglyph analysis
C. Triallel analysis
D. Quadriallel analysis

55. Hull coined the term overdominance in:
A. 1945
B. 1920
C. 1908
D. None

56. Which of the following part of epistasis controlled heterosis?
A. Additive × additive
B. Additive × dominance
C. Dominance × dominance
D. None

57. Who suggested the significance of D square technique for the assessment of genetic diversity in crop plants in 1952?
A. Rao
B. Wright
C. Brim
D. Golden

58. Who developed the semigraphical technique (metroglyph) for identification of phenotypic variability in population?
A. Anderson (1957)
B. Rao (1952)
C. Smith (1936)
D. None

59. Which measures the degree and direction of relationship between two or more variables?
A. Correlation
B. Regression
C. t test
D. None

60. The intensity of cause and effect can be measured among traits by statistics:

A. Correlation coefficient B. Regression coefficient

C. Discriminant function index D. None

61. Which measures used for magnitude change in dependant variable by unit change in the independent variable?

A. Regression coefficient B. Correlation

C. Path coefficient D. All of these

62. Which measures provide the coefficient of determinant (R^2)?

A. Multiple correlation B. Simple correlation

C. Intra-class correlation D. All of these

63. Who developed the concept of path analysis in 1921?

A. Wright B. Dewey

C. Smith D. Hanson

64. Which of the following is the assumption for correlation and path analysis?

A. Linearity B. Additivity

C. Both A and B D. None

65. Which analysis provides the way to know the residual effect?

A. Path analysis B. Correlation coefficient

C. Diallel analysis D. All of these

66. Who proposed the discriminant function for plant selection and advised that a better way of exploiting genetic correlation with several traits having high heritability in 1936?

A. Smith B. Hanson

C. Johnson D. Kempthorne

67. The linear relationship of several traits with yield is called:

A. Selection index B. Heritability

C. Genetic gain D. Selection differential

68. Which of the following is correctly matched?

a.	Classical selection index	*i.*	Wright (1921)
b.	General selection index	*ii.*	Kempthorne and Nordskog (1959)
c.	Restricted selection index	*iii.*	Hanson and Johnson (1957)
d.	Path analysis	*iv.*	Smith (1936)

A. a-iv, b-iii, c-ii, d-i B. a-i, b-ii, c-iii, d-iv

C. a-ii, b-iii, c-i, d-iv D. a-iii, b-ii, c-iv, d- i

69. Who was selection index in soybean for improvement of oil yield in 1959?

A. Brim *et al.* B. Smith

C. Hanson D. Johnson

70. Which genetic assumption is associated with discriminant function or selection index?
    A. Diploid segregation and lack of maternal effect
    B. Absence of linkage, epistasis and multiple allele
    C. Equal survival of all genotypes in population
    D. All of these

71. The diallel mating analysis for polygenic traits in self pollinated crops developed in 1954 by:
    A. Jinks   B. Hayman
    C. Both A & B   D. None

72. Which is associated with diallel mating design?
    A. Crossing of parents in all possible combinations
    B. Diallel includes also one way and reciprocal crosses
    C. Total number of full diallel crosses is equaled to n(n-1)
    D. All of these

73. The material includes for evaluation in diallel analysis:
    A. Parents   B. Reciprocal
    C. One way crosses   D. All of these

74. How many parents can be evaluated by diallel analysis at a time?
    A. 10-12   B. 14-16
    C. 18-20   D. None

75. Half diallel analysis is recommended when material represent:
    A. Parents do not have male sterility or self incompatibility
    B. Reciprocal differences are not significant
    C. Traits show Mendelian inheritance
    D. All of these

76. Which genetic assumption is associated with diallel and partial diallel analysis?
    A. Normal diploid segregation and homozygosity of parents
    B. Random mating and lack of maternal effect
    C. Absence of linkage, multiple alleles and epistasis
    D. All of these

77. How many approaches are used for diallel analysis?
    A. 3   B. 2
    C. 4   D. 7

78. Graphical approach of diallel is based on:
    A. Estimation of component of genetic variation
    B. Estimates of combining ability and effect
    C. Both A & B
    D. None

79. Jinks and Hayman developed which of the following approach of diallel in 1953:

A. Graphical approach
B. Numerical approach
C. Both A & B
D. None

80. Which approach of diallel helps in identification of superior parent?

A. Graphical approach of diallel
B. Numerical approach of diallel
C. Both A & B
D. None

81. Which of the following method of numerical approach of diallel is commonly used for estimation of combining ability ?

A. Method 1
B. Method 2
C. Method 3
D. Method 4

82. In diallel analysis, additive genetic variance is equaled to:

A. Twice of general combining ability variance
B. General combining ability variance
C. Twice of specific combining ability variance
D. Specific combining ability variance

83. In diallel, specific combining ability is equal to:

A. Dominance × Dominance variance
B. Additive × Dominance variance
C. Dominance variance
D. All of these

84. Who developed the concept of partial diallel analysis in 1957?

A. Kempthorne
B. Smith
C. Hanson
D. Hayman and Jinks

85. The total number of crosses will be made in partial diallel mating design equal to

A. ns/2
B. ns/4
C. ns/8
D. None

86. Partial diallel permits the evaluation of maximum inbreds/pure lines at a time:

A. upto 20
B. upto 60
C. upto40
D. None

87. In partial diallel, total number of crosses will be when inbred = 20 and s = 11

A. 110
B. 380
C. 190
D. None

88. Which information can not be obtained from partial diallel analysis?

A. *sca effect*
B. *gca effect*
C. D and H component of variance
D. None

89. Which approach is based on first and second order statistics?

A. Diallel analysis
B. Partial diallel analysis
C. Line x tester analysis
D. All of these

90. In line x tester analysis, additive genetic variance is represented by:
   A. gca variance
   B. sca variance
   C. Additive × dominance variance
   D. None
91. Maximum inbreds can be evaluated by L x T analysis are:
   A. Upto 50
   B. Upto 100
   C. Upto 60
   D. None
92. Line x Tester analysis is the modified form of which cross:
   A. Top cross
   B. Double cross
   C. Partial diallel
   D. None
93. Line x tester provides information about the following:
   A. *gca* and *sca* variance and *gca* effects
   B. D and H component of variance
   C. Both A & B
   D. None
94. Who developed the concept of triallel and quadriallel analysis in 1962?
   A. Rowling and Cokerham
   B. Kempthorne
   C. Lerner
   D. Smith
95. Which technique produces maximum number of crosses remaining equal parents?
   A. Diallel mating design
   B. Triallel mating design
   C. Quadriallel mating design
   D. Line x tester mating design
96. Total number of crosses are produced among *n* inbreds in quadriallel mating design:
   A. $n(n-1)(n-2)(n-3)/8$
   B. $(n-1)(n-2)(n-3)/8$
   C. $(n-1)(n-2)(n-3)/2$
   D. $n(n-1)(n-2)(n-3)/2$
97. Total number of crosses are produced of *n* inbreds in triallel mating design:
   A. $n(n-1)(n-2)(n-3)/8$
   B. $(n-1)(n-2)(n-3)/8$
   C. $(n-1)(n-2)(n-3)/2$
   D. $n(n-1)(n-2)/2$
98. Minimum number of crosses produced among *n* parents by which of the following design:
   A. Diallel mating design
   B. Triallel mating design
   C. Quadriallel mating design
   D. Partial diallel mating design
99. Quadriallel mating design is a modified form of :
   A. Single cross
   B. Top cross
   C. Three way cross
   D. Double cross
100. Triallel mating design is a modified form of :
   A. Single cross
   B. Top cross
   C. Three way cross
   D. Double cross
101. Which technique involves crosses of four different parents?
   A. Diallel mating design
   B. Triallel mating design
   C. Quadriallel mating design
   D. Partial diallel mating design

102. Covariance between parent and offspring is equal to:

A. Half of the additive genetic variance

B. One fourth of the dominance variance

C. One fourth of the dominance variance plus half of the additive genetic variance

D. None

103. Covariance among full sibs is equal to:

A. Half of the additive genetic variance

B. One fourth of the dominance variance

C. One fourth of the dominance variance plus half of the additive genetic variance

D. None

104. One fourth of the dominance variance is :

A. Covariance among half sibs B. Covariance between parent and offspring

C. Covariance among full sibs D. None

105. Which technique of the followings is based on first order statistics?

A. Genetic advance B. Stability

C. Generation mean analysis D. Both B & C

106. The concept of generation mean analysis was developed by:

A. Hayman (1958) B. Jinks (1958)

C. Both A & B D. None

107. Three parameter model of generation mean analysis provides information about:

A. m B. d

C. h D. All of these

108. Who developed three parameter model of generation mean analysis in 1958?

A. Jinks and Jones B. Mather

C. Cavalli D. Kearsey

109. Which model of generation mean analysis is used when the material does not have back crosses but F_3 generation is available ?

A. Five parameter model B. Three parameter model

C. Six parameter model D. All of these

110. Which is truly related with generation mean analysis ?

A. Free from genetic assumptions

B. Combining ability is not measured

C. Provide the information about D, H, and epistatic component of variance

D. All of these

111. Which design requires less area for evaluation of materials ?

A. NCD 3 B. NCD 2

C. NCD 1 D. None

112. The concept of biparental mating design was developed by:
   A. Comstock and Robinson B. Hanson and Smith
   C. Kempthorne D. None

113. Which design is extension of NCD 3:
   A. TTC mating design B. Triallel mating design
   C. Diallel mating design D. None

114. Second order statistics based technique is:
   A. TTC B. Biparental mating
   C. Both A & B D. None

115. Which techniques are used for detection of additive, dominance and epistatic variance in the experimental materials?
   A. Generation mean analysis B. Quadriallel analysis
   C. Triallel analysis D. All of these

116. Both additive and dominance genetic variance can be obtained by which of the following techniques:
   A. Diallel and partial diallel analysis B. Line x tester analysis
   C. Biparental analysis D. All of these

117. The concept of triple test cross is developed by:
   A. Kearsey and Jinks (1968) B. Kempthorne (1958)
   C. Hayman and Jinks (1953) D. None

118. The suitability of a genotype /variety for general cultivation across the environments is known as:
   A. Stability B. Adaption
   C. Adaptability D. None

119. The capacity/ ability of a variety or genotype to exhibit relatively stable performance in different environment is called :
   A. Adaptation B. Stability
   C. Adaptability D. None

120. The fitness of a genotype or population in a particular environment is termed as:
   A. Adaptation B. Adaptability
   C. Both A & B D. None

121. The term co-adaptation was proposed by:
   A. Dobzhansky B. Lerner
   C. Johnson D. None

122. Who introduced the concept of suttle breeding for wheat improvement in 1945?
   A. Norman E. Borlaug B. Warner
   C. Jinks and Jones D. Morgan

123. Which is not correctly matched ?

A. Finlay and Wilkinson model of stability- 1963

B. Eberhart and Russell Model of stability- 1966

C. Perkins and Jinks model of stability - 1968

D. None

124. Accordingly Eberhart and Russell Model of stability (1966), regression coefficient is equal to one results the line/variety is

A. Stable B. Average stable

C. Below average stable D. None

125. Accordingly Finlay and Wilkinson model of stability (1963), regression coefficient is equal to one results the line/variety is

A. Stable B. Average stable

C. Below average stable D. Absolute stable

126. According to Fisher (1918), genetic variances is classed into:

A. Additive genetic variance B. Dominance genetic variance

C. Epistatic variance D. All of these

127. Additive genetic variance is associated with which of the following:

A. Additive gene action B. Homozygosity

C. Polygenic traits D. All of these

128. Which measure is used for obtaining the breeding value ?

A. Additive gene action B. Dominance gene action

C. Both A & B D. Dominance × Dominance variance

129. Which characteristic is not related to dominance genetic variance ?

A. Dominance variance is always lesser than additive genetic variance

B. Positively correlated with specific combining ability, heterozygosiy and found in cross pollinated crop

C. Chief cause of heterosis in crop and exhibit not fixable feature in individual

D. None

130. Higher value of gac than sca indicates of preponderance of

A. Additive gene B. Non additive gene

C. Both A & B D. None

131. The deviation of heterozygotes from mean value due to non allelic interaction is known as:

A. Epistatic genetic variance B. Additive genetic variance

C. Dominance variance D. None

132. The heritable portion of phenotypic variance is known as :

A. Genetic variance B. Environment variance

C. Total variance D. None

133. Accordingly Mather (1949), genetic variances is:
A. Heritable-fixable
B. Heritable-non-fixable
C. Both A & B
D. None

134. Gene action is measured in terms of:
A. Component of genetic variance or combining ability variances and effects
B. Heritability
C. Both A & B
D. None

135. Who suggested combining ability as a measure of various types of gene action in 1942?
A. Sprague and Tatum
B. Lederberg
C. Correns
D. Smith

136. A good index of the transmission of traits from parent to its offspring is called :
A. Genetic gain
B. Genetic advance
C. Heritability
D. None

137. Which factor affects the genetic advance?
A. Heritability
B. Selection intensity
C. Genetic variability
D. All of these

138. Upgrading in the average genotypic value of selected plants over the parental population is called:
A. Genetic advance
B. Heritability
C. Genetic gain
D. Selection differential

139. Genetic gain is denoted by which symbol:
A. R
B. B
C. r
D. b

140. Which is not correct for breeding value:
A. Provides the idea about transmits ability of the parents
B. It is associated with additive gene action
C. It is also associated with *gca*
D. None

141. The value transmitted from parent to its progeny is called:
A. Breeding value
B. General combining ability
C. Heritability
D. None

142. Which measures are associated with breeding value in crops plant?
A. Prepotency, heritability and parent offspring regression
B. Combining ability and mean deviation
C. Genetic advance
D. All of these

143. Narrow sense heritability is associated with:
A. Additive genetic variance
B. Breeding value
C. Additive gene effect
D. All of these

Terms

- **Adaptability:** Ability of a variety or genotypes to exhibit relatively stable performance in different environments.
- **Adaptation:** A feature or performance or fitness of genotypes seem in a particular environmental condition.
- **Additive variance:** The variance created by average effects of genes on all segregating loci, it measures additive gene action.
- **Biparental cross mating:** A plan in which plant are randomly selected from F_2 or more advanced generation of a cross among two pure lines have dissimilar features and the selected plant are cross in a definite scheme.
- **Breeding value:** It is the value transmitted from parent to their offspring. It is vastly used in animal breeding and include both additive and nonadditive gene action.
- **Combining ability:** Ability of a genotype to transmit the superior performance to its progeny.
- **Convergence:** The structure is similar between the two species because of independent adaptation to common environmental pressures.
- **Correlation coefficient:** Numerical value of correlation.
- **Correlation:** The linear relationship between two or more variable and their range lies between 0-1.
- **D square analysis:** A statistical method, which is used for estimation of genetic diversity within the population. According this analysis, similar genotypes are grouped in same cluster.
- **Diallel Analysis:** A method of analysis which is used to estimate different genetic component of variances in dialle crosses.
- **Diallel mating:** A procedure in which selected 'n' lines cross each other in all possible combinations. Total number of one way cross is equal to n (n-1)/2.
- **Dominance variance:** Variance caused by intra allelic interaction. It is a measure of dominance gene action.
- **Environmental variation:** The variation caused by nature in population.
- **Epistatic variance:** It includes both additive and non-additive components of genetic variance.
- **Full diallel mating design:** A crossing plan in which crosses are made between selected plants in all possible combinations (including reciprocal cross).

- **General combining ability:** Average performance of a genotypes in a number of crosses when cross have one common parent.
- **Generation mean analysis**: It provide the estimate of genetic component of variances of six different generation of a cross *viz.*, parent (P_1 and P_2), their F_1, F_2 and backcrosses (B_1 and B_2). It has three models to be used for estimation of genetic component of variances *viz.*, six parameter model, five parameter model and three parameter model.
- **Genetic advance:** Advancement in the average genotypic value of selected plants over the parental population.
- **Genetic variance:** The sum additive and non-additive genetic variance (VA and VD).
- **Genetic variation:** Variation caused due to differences in genetic constitution of individuals.
- **Half diallel mating:** In which all possible crosses between selected "n" lines are made in one direction only.
- **Heritability:** Capacity of traits to transmit from parents to their offspring. In other words, the genetic variation is heritable because it can be passed from parent to offspring.
- **Heterosis:** Superior performance of F_1 compared to their both parents.
- **Line × Tester mating:** It is a modifying form of top cross in which some selected plant is denoted as male and another female for making crosses. A series of crosses in which one parent is common known as tester. The total number of crosses is equal to L (female lines) × T (male lines).
- **Mean:** The value obtained by dividing the sum total of all observation by number of observations.
- **Metroglyph analysis:** Method, which is used to estimate phenotypic variability present within population. It is a semi graphical approach.
- **Narrow sense heritability:** The proportion of the phenotypic variation which is due to additive genetic effects.
- **Parameter:** Numerical quantity which describes some characteristic of a population.
- **Partial diallel mating:** It is a modifying form of diallel mating in which only a part of all possible combinations are made between selected plants and utilized for evaluation. Total number of crosses is equal to ns/2.
- **Phenotypic covariance:** The product obtained by dividing genotypic covariance with environmental covariance.
- **Phenotypic variation:** The sum total of genetic and environmental variation.
- **Quadriallel mating:** It is the extension of double crosses and produce total number of crosses is equal to n (n-1) (n-2) (n-3)/8.
- **Quantitative feature:** A measurable morphological characteristics.

- **Quantitative genetics:** The study of inheritance of minor or quantitative genes or characters with their variation.
- **Regression coefficient:** It is a measure to quantify the rate of change in dependant variable for unit change in the independent variable.
- **Selection differential:** The differences between the mean phenotypic value of selected plants and mean phenotypic value of parental population. It is denoted by K.
- **Selection index:** The linear relationship of several traits with yield.
- **Selection intensity:** The proportion of the plants selected to grow in the next generation.
- **Sibbing:** The inter-mating among the individuals of the same parents.
- **Specific combining ability:** The deviation in performance of a cross from that predicted on the basis of general combining ability.
- **Stability:** Suitability of genes or genotypes to produce average performance in wide range of environmental condition.
- **Triallel mating:** It is the extension of three way cross and produce total number of cross is equal to n (n-1) (n-2)/2.

ANSWERS

Chapter 8: Quantitative genetics

1. A	31. B	61. A	91. A	121. A
2. A	32. A	62. A	92. A	122. A
3. A	33. B	63. A	93. C	123. D
4. A	34. B	64. C	94. A	124. A
5. B	35. B	65. A	95. C	125. B
6. A	36. D	66. A	96. A	126. D
7. B	37. A	67. A	97. D	127. D
8. D	38. A	68. A	98. D	128. A
9. C	39. A	69. A	99. D	129. D
10. A	40. D	70. D	100. C	130. A
11. A	41. A	71. C	101. C	131. A
12. B	42. A	72. D	102. A	132. A
13. A	43. A	73. D	103. C	133. C
14. B	44. A	74. A	104. A	134. C
15. A	45. A	75. D	105. D	135. A
16. A	46. D	76. D	106. C	136. C
17. C	47. D	77. B	107. D	137. D
18. A	48. A	78. A	108. A	138. A
19. A	49. C	79. A	109. A	139. A
20. D	50. B	80. A	110. D	140. D
21. A	51. A	81. B	111. A	141. A
22. A	52. A	82. A	112. A	142. D
23. A	53. C	83. C	113. A	143. D
24. D	54. B	84. A	114. C	
25. D	55. A	85. A	115. D	
26. A	56. C	86. A	116. D	
27. A	57. A	87. A	117. A	
28. A	58. A	88. A	118. A	
29. A	59. A	89. D	119. C	
30. A	60. B	90. A	120. A	

Chapter 9

Chromosomal Aberrations

1. How many types of aberrations are found in chromosome?
   A. Structural aberration of chromosome
   B. Numerical aberration of chromosome
   C. Both A & B
   D. None
2. A complete set of basic chromosome is known as :
   A. Genome B. Proteomes
   C. Haploid D. None
3. Natural variation in number of chromosomes or morphological change of chromosome in respect of normal diploid individual is called :
   A. Changes of chromosome structure B. Chromosome aberration
   C. Chromosomal aberration in number D. All of these
4. Somatic cell of an individual carries :
   A. Two copy of each chromosome of the genome of a species.
   B. Only one chromosome number of each homologous chromosomes pair
   C. A single copy of genomic chromosomes
   D. Both A & C
5. Which of the following aberration, is considered as structural aberration of chromosome?
   A. Deletion or deficiency B. Duplication
   C. Translocation D. All of these
6. Which of the following case is involved in inversion?
   A. Rotated segments of chromosomes at 180°
   B. Rotated gene sequence
   C. Acentric chromosome produced through structural changes
   D. All of these

7. Structural aberration of chromosome is caused by :
   A. Chromosome breakage B. Natural mutation
   C. Induced mutation D. All of these
8. Which factors leads to change in the frequency of spontaneous chromosome breakage?
   A. Temperature
   B. Age of individual
   C. Oxygen availability and metabolic activity of cell
   D. All of these
9. Which of the following enzymes is responsible to digest acentric chromosome?
   A. Exonuclease B. Protease
   C. Helicase D. None
10. Variation observed in the structure of chromosome, is called :
   A. Morphological change in chromosome B. Structural changes in chromosome
   C. Both A & B D. None
11. Translocation involves exchange of chromosomal segments between one of the following:
   A. Homologous chromosomes B. Sister chromatid of a chromosome
   C. Non homologous chromosome D. All of these
12. Deletion refers to :
   A. Lack of chromosomal segment B. Addition of chromosomal segment
   C. Deficiency of chromosomal segment D. Both A & C
13. Duplication refers to :
   A. Multiplication of a chromosomal segment
   B. Duplication of a chromosomal segment
   C. Repeatability of chromosomal segment
   D. All of these
14. Progeny of translocated chromosomes may be
   A. Structural homozygotes B. Structural heterozygotes
   C. Both A & B D. None
15. Loss of a part of chromosome is considered as :
   A. Deficiency or deletion B. Translocation
   C. Repeats D. Inversion
16. A recessive allele will behave like a dominant allele due to :
   A. Repeats B. Translocation
   C. Deletion D. Inversion
17. A recessive allele behaves as a dominant allele in hemizygous condition is called:
   A. Dominance B. Pseudo-dominance
   C. Recessive D. Co-dominance

18. Deficiency is found in :

A. *Drosophilla* B. Maize

C. Tomato D. All of these

19. Duplication occurs only in :

A. One chromosome of the two homologous chromosome

B. Both homologous chromosomes

C. Two non sister chromatids of a chromosome

D. None

20. Duplication occurs at :

A. Pachytene phase of prophase in meiosis

B. Metaphase of mitotic division

C. Anaphase

D. Interphase

21. Duplications are as :

A. Tandem repeats B. Reverse tandem repeats

C. Displaced repeats D. All of these

22. Bar trait arises in *Drosophilla* at the region of 16 A of X chromosomes due to :

A. Deletion B. Translocation

C. Duplication D. Inversion

23. Progeny of Bar eye *Drosophilla* has ultra bar and normal wild type eyes individuals due to :

A. Translocation B. Inversion

C. Unequal crossing over D. Deletion

24. Translocation is the result of :

A. Exchange of chromosomal segment between non homologous chromosome

B. Exchange of chromosomal segment between a pairs of homologous chromosome

C. Deletion of chromosomal segment

D. None

25. Mutual exchange of chromosomal segments between two pairs of non homologous chromosome is considered as :

A. Reciprocal translocation B. Segmental interchange translocation

C. Both A & B D. None

26. Normal pairing of homologous chromosomes is not possible in one of the following situation:

A. Duplication B. Translocation heterozygote

C. Translocation homozygote D. None

27. Pairing between homologous chromosomes can be observed at :
    A. Pachytene phase
    B. Zygotene
    C. Metaphase
    D. All of these
28. First of all natural translocation was observed in :
    A. *Oenothera*
    B. Sweet pea
    C. Garden pea
    D. All of them
29. First of all spontaneous translocation was reported by :
    A. De vries
    B. Morgan
    C. Camillo
    D. Kumar, B N
30. Presence of translocation heterozygosity is determined with the help of :
    A. Presence of semi-sterility
    B. Low seed set
    C. Fully fertile progeny
    D. Both A & B
31. Which types of progeny is produced by a heterozygote translocated plant?
    A. One translocated heterozygote plant
    B. Two normal plant
    C. One translocated homozygote plant
    D. All of these
32. Translocation heterozygotes produce:
    A. Fully fertile progeny
    B. Fully sterile progeny
    C. Partial sterile progeny
    D. None
33. Which of the following structural changes includes two breaks in a chromosome and their intercalary segment reunites in reverse order ?
    A. Duplication
    B. Inversion
    C. Both A & B
    D. Translocation
34. Inversion may be one of the following :
    A. Paracentric type
    B. Pericentric type
    C. Both A & B
    D. None
35. Which of the following events occur due to pericentric inversion?
    A. Inverted segment contains centromere
    B. Inverted segment do not contain centromere
    C. Change the morphology of chromosome by shifting the position of centromere
    D. Both A & C
36. Centromere does not involve in which of the following types of inversion?
    A. Paracentric inversion
    B. Paricentric inversion
    C. Both A & B
    D. None
37. Paracentric inversion does not alter :
    A. Morphology of chromosome
    B. Number of genes
    C. Position of centromere
    D. All of these

38. Pericentric inversion alters :
A. Position of centromere
B. Morphology of chromosome
C. Number of gene
D. All of these

39. Odd number of crossovers in inverted region results into the formation of:
A. Acentric chromosome
B. Dicentric chromosome
C. Chromosome without centromere
D. All of these

40. Dicentric and acentric chromatids are observed, in which of the following phase of cell division?
A. Anaphase I
B. Prophase I
C. Metaphase II
D. None

41. Double cross over may produce:
A. Deficiency
B. Duplication
C. Deletion
D. All of these

42. Deficiency is known as:
A. Deletion
B. Inversion
C. Duplication
D. None

43. Paracentric inversion do not produce, which of the following events?
A. Dicentric bridge
B. Acentric fragment
C. Both A & B
D. None

44. Inversion is considered to be as:
A. Crossover suppressor
B. Crossover enhancer
C. Crossover initiator
D. None

45. Reduction of crossover results :
A. Less recovery of the products of single crossovers
B. More recovery of the products of single crossovers
C. Parental type products only
D. None

46. Who studied the inversion effect in the production of *CIB* stock?
A. Knight
B. Mendel
C. Morgan
D. H. J. Muller

47. *CIB* stock is used for :
A. Detection of sex linked lethal mutation
B. Detection of chromosomal aberration
C. Detection of inversion
D. All of these

48. When a secondary inversion occurs in region of a chromosome, which has already inverted sequence then it is called :

A. Overlapping inversion
B. Paracentric inversion
C. Pericentric inversion
D. None

49. Who studied the relationship between races on the basis of overlapping inversion in trees?

A. Th. Dobzhansky
B. A. H. Sturtevant
C. Both A & B
D. None

50. Which of the following types of inversion has considerable partial male sterility?

A. Inversion heterozygote
B. Inversion homozygote
C. Both A & B
D. Overlapping inversion

51. Who was the first to invent inversion in *Drosophila melanogester?*

A. Sturtevant, 1900
B. Sturtevant and Plunkett, 1926
C. Bridges, 1937
D. Muller, 1898

52. Presence of inversion in a chromosome can be detected using :

A. Phenotypic expression
B. Genetic expression
C. Cytological expression
D. All of these

53. Breakage fusion bridge cycle occurrance in a duplication heterozygote of maize during anaphase was reported by:

A. McClintock
B. Joseph, J. K.
C. Sturtevant
D. Punnett

54. An iso-chromosome consist of :

A. Two identical arms in relation their morphology and gene constitution
B. Two non identical arm
C. Two identical arm in their morphology
D. None

55. Crossing over is absent in :

A. Deleted region of chromosome
B. Male *Drosophilla*
C. Both A & B
D. None

56. In *Drosophilla*, dicentric bridge at anaphase I is associated with one of the following events:

A. Movement of crossover products
B. Create male sterility
C. Chromosomes will not be incorporated into either progeny cell
D. All of these

57. Breakage of DNA at two different locations and reunion of the broken ends resulting a new arrangement is caused by:

A. Deletions
B. Duplications
C. Inversions and translocations
D. All of these

58. Which of the following events, occurs in the production of a new rearrangement of genes:

A. Generation of two or more double-strand breaks in the chromosomes of a cell

B. Broken (non-telomeric) ends of chromosome cannot replicate properly

C. Both A & B

D. None

59. Analysis of inversions are carried out in cells that contain one normal haploid chromosome set alongwith one set carrying the inversion, such cells are called :

A. Inversion heterozygote
B. Inversion homozygote
C. Tandem repeats duplication
D. All of these

60. Which of the following events does not observed during paracentric inversion ?

A. No any dicentric bridge
B. No any acentric chromosomal ferments
C. Both A & B
D. None

61. During inversion, a two strand double cross over produces, which of the following ?

A. Two normal chromatid
B. Two inverted chromatid
C. Both A & B
D. None

62. During inversion, a three strand double cross over produces, which of the following ?

A. An acentric fragment of chromosome
B. A dicentric chromosome
C. Both A & B
D. None

63. During inversion, a four strand double cross over produces, which of the following ?

A. Two acentric fragment of chromosome
B. Two dicentric chromosome
C. Both A & B
D. None

64. Shift type of translocation is naturally found in :

A. *Drosophilla*
B. *Neurosphora*
C. Both A & B
D. None

65. Which of the following properties carries by inversion ?

A. Formation of dicentric bridge
B. Formation of acentric fragment
C. Rotation of chromosome at 180°
D. All of these

66. Which of the following symbol represent of crossover suppressors ?

A. C
B. D
C. N
D. K

67. *Oenothera races* involve about how many non-homologous chromosomes in translocation:

A. Two to seven non-homologous chromosome

B. Thirteen to fifteen homologous chromosome

C. All homologous chromosome

D. None

68. Maximum non-homologous chromosomes cause reciprocal translocation in *Oenothera lamarckiana* :

A. 6 B. 21
C. 5 D. 8

69. If a chromosome have deletion results in :

A. Chromosome thick than normal chromosome
B. Chromosome longer than normal chromosome
C. Chromosome short than normal chromosome
D. No changes in chromosome

70. Which of the following changes are found in chromosome resulting duplication ?

A. Length of chromosome is more than normal chromosome
B. Length of chromosome is less than normal chromosome
C. No change in length in respect to normal chromosome
D. None

71. Misdivision of centromere results :

A. Production of normal chromosome B. Production of telocentric chromosome
C. Production of acentric chromosome D. Production of metacentric chromosome

72. Variation in number of chromosomes is referred to as :

A. Euploidy B. Aneuploidy
C. Both A & B D. None

73. Either addition or deletion of chromosomes lead to the development of :

A. Nullisomic B. Monosomics
C. Trisomics D. All of these

74. Which of the following conditions arises when individual lack in number of chromosomal compliments ?

A. Nullisomy B. Monosomy
C. Both A & B D. None

75. Which of the following conditions arises due to addition of chromosomal compliments in an individual ?

A. Trisomy B. Tetrasomy
C. Both A & B D. All of these

76. A condition arises due to lack of one chromosome in diploid :

A. Simple monosomy B. Nullisomic
C. Double monosomy D. Tertiary monosomy

77. How many monosomics are possible in bread wheat ?

A. 44 B. 21
C. 43 D. 42

78. Which of the following symbols represents monosomy condition ?

A. 2n – 1
B. 2n + 1
C. 2n – 1 + 1
D. All of these

79. Which of the followings, shows aneuploidy condition of an organism ?

A. 2n–1 and 2n – 2
B. 2n + 1 and 2n + 2
C. 2n–1–1–1
D. All of these

80. What is the symbol to be used for nullisomy ?

A. 2n – 2
B. 2n – 1
C. 2n + 2
D. 2n + 1

81. Pentasomy is a condition of an individual includes :

A. Tetrasomics
B. Monosomics
C. Nullisomics
D. All of these

82. Trisomy causes, which of the following diseases ?

A. Down syndrome
B. Edwards syndrome
C. Peatau syndrome
D. All of these

83. Which of the following methods is used to detect Germline aneuploidy ?

A. Karyotyping
B. Chromosomal banding
C. Both A & B
D. None

84. Who isolated the 21 monosomics in wheat (Chinese spring type wheat) crop ?

A. Sears, E. R.
B. Muller
C. Jensen
D. Koelreuter

85. Who isolated monosomics in cotton (*Gossipium hirsutum* L., 2n = 52) ?

A. E. R. Sears et al.
B. J. E. Endrizzi *et al.*
C. E. R. Clausen
D. None

86. Which of the following conditions reduces size, vigour and fertility of plant compared to normal plant ?

A. Nullisomy
B. Monosomy
C. Aneuploidy
D. All of these

87. Which of the following formula, shows triple monosomics ?

A. 2n – 1 – 1 – 1
B. 2n + 1 + 1 + 1
C. 2n – 1 + 1 + 1
D. None

88. Which of the following events occurs when a plant loose of different chromosomes ?

A. Produce different morphology of plant population
B. Either increase or loss of vigour of plant population
C. Change in numbers of chromosomes of plants
D. All of these

89. All possible number of nullisomics present in wheat (2n = 42) :

A. 21 B. 34
C. 20 D. 80

90. Double monosomics is caused by which of the following ?

A. Condition arises due to missing of two non-homologous chromosomes
B. Condition arises due to missing of two homologous chromosomes
C. Condition arises due to addition of two non-homologous chromosomes
D. Condition arises due to missing of three non-homologous chromosomes

91. All possible number of trisomics is always equal to:

A. Haploid chromosomes number B. Somatic chromosome number
C. Diploid chromosome number D. Disomy chromosome number

92. Symbol of disomics is:

A. 2n B. 2x
C. Both A & B D. None

93. Which of the following individuals has an isochromosome in the form of extra chromosome?

A. Secondary trisomics B. Primary trisomics
C. Nullisomics D. All of these

94. Identical extra chromosome carried by which of the following individuals ?

A. Double trisomics B. Primary trisomics
C. Nullisomics D. All of these

95. Which of the following individuals carries an extra chromosome translocated ?

A. Double trisomics B. Primary trisomics
C. Tertiary trisomics D. All of these

96. Who was the first to work on Jimson weed to find out the trisomy ?

A. A. F. Blakslee *et al.* B. E. R. Clausen and D. R. Cameron
C. E. R. Sears et al. D. None

97. E. R. Sears was the brilliant scientist, who died working on wheat in ?

A. 1991 B. 2005
C. 2009 D. 2006

98. *T. Tsuchiya* passed away in :

A. May, 1992 B. Aug, 1999
C. Jan, 2012 D. May, 1988

99. Trisomy is found in :

A. *Humo sapiens* B. *Hordium vulgare*
C. *Datura stramonium* D. All of these

100. Which of the following characteristics is associated to Mongolism (Down's syndrome) ?

A. Mental retardation
B. A short body
C. Swollen tongue
D. All of these

101. Which of the following gametes is responsible to create trisomy condition in offspring in nature ?

A. n + 1
B. 2n + 1
C. N – 1
D. X – 1 + 1

102. Robertsonian translocation was 1st described by which of the following in grasshoppers in 1916 ?

A. William Rees Brebner Robertson
B. Sturtevant
C. Mc Cliktock
D. None

103. Robertsonian translocation is also known as :

A. Centric-fusion translocations
B. Whole-arm translocations
C. Both A & B
D. None

104. Robertsonian translocation occurs in :

A. Two different chromosome
B. A pair of homologous chromosome
C. Both A & B
D. None

105. Robertsonian translocation is a type of :

A. Reciprocal translocation
B. Simple translocation
C. Shift translocation
D. All of these

106. The first World Down Syndrome Day was celebrated :

A. 21 March 2006
B. 21 December 2006
C. 8 March 2006
D. None

107. Addition of a pair of homologous chromosome to compensate the loss of another pair of homologous chromosome, represented by :

A. 2n – 2 + 2
B. 2n – 1 + 1
C. 2n – 1
D. 2n + 2

108. Who synthesized a complete set of compensating nullisomics, tetrasomics chromosome?

A. E. R. Sears
B. E. R. belling
C. Lush
D. Anderson

109. Monosomics has forty four autosomes plus one extra chromosomal complement cause:

A. Turner syndrome
B. Down syndrome
C. Both A & B
D. None

110. Match column A with column B

Column A	*Column B*
a. E. R. Sears	1. Barley
b. L. Down	2. Down syndrome
c. H. H. Turner	3. Wheat
d. Rajhathy	4. Turner syndrome

A. a-1, b-3, c-2, d-4 B. a-4, b-3, c-2, d-1
C. a-3, b-2, c-4, d-1 D. a-2, b3, c1, d-4

111. A plant breeder made crosses between tetrasomic and disomic plants. The progenies of the crosses would show abundant frequencies of :
A. Frequency of trisomics B. Frequency of trisomics
C. Do not produces trisomics progeny D. None

112. Alien substitution refers to:
A. Chromosome of a species replaced by a chromosome of another species
B. Chromosome of a species replaced by chromosome of same species
C. Chromosome of a species pair with chromosome of same species
D. None

113. Alien addition refers to :
A. Chromosome of a species replaced by chromosome of another species
B. Chromosome of a species replaced by chromosome of same species
C. A pair of chromosome addition from another species in individual
D. None

114. Who gave the term alien substitution line in 1938?
A. Kihara et al. B. Kattermann
C. Click D. Watson

115. Who proposed the term "Alien addition" in 1924?
A. Kattermann B. Errera
C. Leighty and Taylor D. None

116. Colchicine is an alkaloid, obtained from which of the following plants ?
A. *Colchicum autumnale* B. *D. stromoneum*
C. *T. dicoccum* D. None

117. Chemical formula of colchicine is :
A. $C_{22}H_{25}O_6$ B. $C_{22}H_{25}O_6N$
C. C_6 D. None

118. Colchicine alkaloid is used for :
A. Doubling of chromosome in plant B. Breaks the spindle fibres during cell division
C. Development of tetraploids D. All of these

119. Colchicine is extracted from which of the following part of *Colchicum autumnale* ?

A. Bulbs
B. Seeds
C. Both A & B
D. None

120. Name the Nobel Prize winner for reporting colchicine effects:

A. Santi and Singh J.
B. Blakeslee and Avery
C. Bubanda Kanuja
D. None

121. Euhaploids are isolated from which of the followings ?

A. Euploids
B. Aneuploids
C. Monosomics
D. Double monosomics

122. Aneuhaploids originate from which of the followings ?

A. Polyploids
B. Euhaploids
C. Aneuploids
D. Hexaploids

123. Spontaneously, haploids may originate by :

A. Parthenogenetic development of egg cell in flowering plant
B. Somatic cell in flowering plant
C. Germ cell of flowering plant
D. None

124. Androgenic haploids originate from:

A. Synergids cell of embryo sac
B. Antipodal cell of embryo sac
C. Both A & B
D. None

125. Method used to develop haploids is :

A. X-rays treatment and pollen culture
B. Delayed pollination and temperature shocks
C. Distant hybridization
D. All of these

126. Which of the formula represent substitution haploids ?

A. $n - 1 + 1$
B. $2n - 1$
C. $n - 1$
D. All of them

127. Polyhaploids represent by which of the following ?

A. $n = 3x = 21$
B. $n = x = 7$
C. Both A & B
D. $2n = x = 10$

128. Technique of anther and pollen culture was developed in 1964 for production of haploids by one of the following:

A. Guha and Maheshwari
B. Guha and Borlaug
C. Maheshwari and Vavilov
D. None

129. Anther culture technique is utilized for the production of haploids in:

A. Rice
B. Wheat
C. Tobacco
D. All of these

130. Morphology of haploids differ from normal plant:

A. Haploids has smaller seed size than normal seed size of plant

B. Haploids has both smaller stomata and pollen size than normal stomata and pollen size plant

C. Haploids has small flower size than normal plant flower

D. Both B & C

131. Three similar genomes are found in a plant. This plant is called :

A. Autotriploids | B. Aututetraploids

C. Poly-hexaploids | D. None

132. A genotype formed, when haploid cells undergo chromosome doubling is known as :

A. Double haploids | B. Haploids

C. Triploids | D. None

133. Genotype developed by mating between tetraploids and diploids known as:

A. Triploids | B. Hexaploids

C. Aneuploids | D. Auto-tetraploids

134. Common example of natural autopolyploidy is:

A. *Cynodon dactylon* | B. *Tritium monococcum*

C. *Pisum sativum* | D. *Cicer arietinum*

135. Autotriploidy condition found in:

A. Tomato | B. Watermelon and sugarbeet

C. Graps and banana | D. All of these

136. Production of allopolyploids is due to:

A. Doubling of chromosome number in an F_1

B. Doubling of chromosome number in an allo triploids

C. Crossing between two different genera

D. All of these

137. Doubling of chromosome of F_1 results in production of :

A. Amphidiploids | B. Allotetrapolids

C. Triploids | D. Both A & B

138. Which of the following conditions shows amphidiploid?

A. AABB | B. AA

C. Both A & B | D. None

139. The concentration of colchicine alkaloid to be used for doubling of chromosomes varies:

A. 0.01% to 0.5% | B. 0.06 % to 1.0%

C. 1.0 % to 2.0% | D. None

140. Leaves, flower, fruits and seeds are larger than to diploid, in which of the following conditions?

A. Autopolyploidy
B. Hypoploids
C. Hyperploids
D. All of these

141. *Chimeras* is associated to :

A. A mixture of genetically diverse tissues in the same organs
B. A grafting product
C. Organ arise due to mixture of two different tissues
D. All of these

142. Which of the following crops carry *chimeras* tissue?

A. Paddy
B. Sorghum spp.
C. Millet
D. All of these

143. Effect on alien pollen seen on the endosperm and embryo is known as :

A. *Xenia effect*
B. Metaxenia effect
C. Both A & B
D. None

144. The effect of alien pollen on maternal tissues of the fruit is known as:

A. *Xenia*
B. *Metaxenia*
C. Both A & B
D. None

145. Who developed *Raphanobrassica* by crossing between *Raphanus sativus* and *Brassica oleraceae* in 1927?

A. Karpechenko, G. D.
B. Kihara, H.
C. McFadden, E.S. and Sears, E.R.
D. Beasley, J.O.

146. In 1946, who synthesized hexaploid (*Triticum spelta*) wheat?

A. Karpechenko, G. D. and Beasley, J.O.
B. Kihara, H. and Yadav, R. K.
C. McFadden, E.S. and Sears, E.R.
D. Beasley, J.O. and Mohammad, P.K.

147. Artificially alloploids have been successfully developed in :

A. Wheat (*Triticum spelta*)
B. Cotton (*Gossypium hirsutum*)
C. Townsend's grass (*Spartina townsendii*)
D. All of these

Terms

- **Alien addition:** A pair of chromosomes extra in an individual from another species.
- **Alien substitution:** Chromosome of a species replaced by a chromosome of another species.
- **Aneuploids variation:** Variation which occurs due to addition or deletion of one or few changes in the number of only single chromosome.
- **Autopolyploids:** Cell or individual having more than two genomes.
- **Autopolyploidy:** Condition which arises by multiplication of chromosome sets.
- **Chimeras:** A single organism composed of genetically distinct tissues occurs due to either colchicine treatment or grafting and is found commonly in plant.
- **Chromatid:** Longitudinal division of chromosome due to duplication of DNA resulting same arms attached with centromere.
- **Chromosomal aberration:** Alteration in chromosomes.
- **Deletion:** It is also known as deficiency and a type of structural change in chromosome.
- **Double haploids:** A genotype formed, when haploid cells undergo chromosome doubling.
- **Double monosomy:** Condition arises due to missing of two non-homologous chromosomes in cell.
- **Duplication:** A process by which segment of chromosome is duplicated.
- **Euploids:** Cells or individuals have chromosome complements consisting of whole sets or genomes and the basic chromosome number represented by haploid (X).
- **Genome:** A complete set of basic chromosome.
- **Inversion heterozygote:** When one of the homologous chromosomes has inverted segment and another homologous chromosome is normal.
- **Iso-chromosome:** Chromosome has two identical arms in relation their morphology and gene constitution.
- **Metaxenia effect:** The effect of alien pollen on maternal tissues of the fruit
- **Nullisomics:** Plants in which a homologous chromosome pair is completely missing.
- **Numerical chromosomal aberration:** Changes occur in number of chromosomes in cell.
- **Polyploids:** Organism which have more than two genomes.
- **Primary trisomy:** Condition arises due to additional and identical extra chromosome complements in cell.

- **Pseudo-dominance:** A recessive allele behaves as a dominant allele in hemizygous condition.
- **Secondary trisomics:** Individual has an isochromosome in the form of extra chromosome.
- **Simple monosomy:** A condition arises due to lack of one chromosome in diploid.
- **Somatic cell:** Cell has two copy of each chromosome of the genome of a species.
- **Structural chromosomal aberration**: Alteration in the structure of chromosome *i.e.*, shape, size, etc. considered as structural chromosomal aberration.
- **Tetraploids:** Cells have four sets of chromosome (4n).
- **Translocation:** Exchange of chromosomal segments between non-homologous chromosomes.
- **Triploids:** Individuals that contains three set of chromosomes. Genotype developed by mating between tetraploids and diploids; most appropriate example of triploids as banana, water melon, sugarabeat, grapes etc.
- **Trisomics:** A cell or organism arises due to addition of single chromosome in an individual (2n + 1).
- **Unequal crossing over:** Crossing over between repeated DNA sequences that have paired out of register, creating duplication and deletion.
- **Xenia effect:** Effect on alien pollen seen on the endosperm and embryo.

ANSWERS

Chapter 9: Chromosomal aberrations

1. C	31. D	61. C	91. A	121. A
2. A	32. C	62. C	92. C	122. C
3. D	33. B	63. C	93. A	123. A
4. A	34. C	64. C	94. B	124. C
5. D	35. D	65. D	95. C	125. D
6. D	36. A	66. A	96. A	126. A
7. D	37. D	67. A	97. A	127. A
8. D	38. D	68. A	98. A	128. A
9. A	39. D	69. C	99. D	129. D
10. C	40. A	70. A	100. D	130. D
11. C	41. D	71. B	101. A	131. A
12. D	42. A	72. C	102. A	132. A
13. D	43. C	73. D	103. C	133. A
14. C	44. A	74. C	104. C	134. A
15. A	45. A	75. C	105. A	135. D
16. C	46. D	76. A	106. A	136. D
17. B	47. D	77. B	107. A	137. D
18. D	48. A	78. A	108. A	138. A
19. A	49. C	79. D	109. A	139. A
20. A	50. A	80. A	110. C	140. A
21. D	51. B	81. A	111. A	141. D
22. C	52. D	82. D	112. A	142. D
23. C	53. A	83. C	113. C	143. A
24. A	54. A	84. A	114. B	144. B
25. C	55. C	85. B	115. C	145. A
26. B	56. D	86. D	116. A	146. C
27. A	57. D	87. A	117. B	147. D
28. A	58. A	88. D	118. D	
29. A	59. A	89. A	119. C	
30. D	60. C	90. A	120. B	

Chapter **10**

Mutation

1. A heritable change in an individual is known as:
   - A. Mutation
   - B. Gene regulation
   - C. Genome
   - D. None
2. Chromosomal mutation differs from point mutation in respect of :
   - A. Change in chromosome structure or number
   - B. Change in gene or change in base of DNA
   - C. Change in somatic cell
   - D. All of these
3. Point mutation is also known as :
   - A. Gene mutation
   - B. DNA mutation
   - C. Both A & B
   - D. Somatic mutation
4. Who coined the term mutation in1901 working on *Oenothera?*
   - A. De Vries
   - B. Morgan
   - C. Wright
   - D. Shull
5. Who reported mutation working on fruitfly (*Drosophilla mela*... 1910?
   - A. T.H. Morgan
   - B. Hugo de Vr...
   - C. Sewall Wright
   - D. None
6. The theory of mutation was based on which of the f...
   - A. *Oenothera lamarckiana*
   - B. T... *marckiana ?*
   - C. *Brassica juncea*
   - D. ... nathan
7. Who put forth the theory of mutation work...
   - A. De Vries
   - C. Bridges

8. In which of the following, mutation was observed ?

A. A male sheep breed
B. A female sheep
C. A male *Drosophilla*
D. A female *Drosophilla*

9. Which of the following sheep breed was developed using short legged lamb?

A. Ancon
B. Apennine
C. Arabi
D. Arapawa Island

10. Who observed a short legs lamb in a flock of sheep in 1791 ?

A. Seth Wright
B. Sewall Wright
C. Muller
D. Morgan

11. Mutation occurs in :

A. Gametes
B. Zygote
C. Chromosome and DNA
D. All of these

12. High frequencies of mutation occurs in:

A. Germ cell
B. Gametes
C. Both A & B
D. Somatic cell

13. Mutation is found in :

A. Eukaryotic cell
B. Prokaryotic cell
C. Both A & B
D. None

14. Who used X-rays first of all to induce mutation in crop plant ?

A. Stadler
B. Muller
C. Morgan
D. None

'ho reported mutagenic effect of X-rays on *Drosophilla* ?

Muller
B. Johnson
16. 'sen
D. Stadler

spontaneous mutation varies :

C. 10^{-7}
B. 10^4 to 10^8

17. Wh B
D. None

A. rays based on cathode radiation in 1895?

C. St en
B. B. McClintock

18. Exampl
D. None

A. Epilc

C. Both tion is :

B. Notch wings in *Drosophilla*

D. None

19. Mutator genes leads to :

A. Increase spontaneous mutation rate in their neighboring genes

B. Decrease spontaneous mutation rate in their neighboring genes

C. Artificial matable genes

D. All of these

20. Which of the following, factor prevents mutation ?

A. Mutagens B. Antimutator genes

C. Mutable genes D. All of these

21. Which of the following factors or agents are used to enhance the induce mutation?

A. Chemical mutagens B. Physical mutagens

C. Both A & B D. None

22. Induced mutation originate by using :

A. X rays treatment B. 5-BU treatment

C. 2-AP treatment D. All of these

23. A site of gene which has high mutation rate is called :

A. Hot spot B. Cool spot

C. Antimutable spot D. All of these

24. Who discovered white eye mutant in *Drosophilla* in 1910?

A. T. H. Morgan B. McClintock

C. Both A & B D. K. C. Mehta

25. Mutation found in :

A. Pea B. Maize and Snapdragon

C. Rodents and Human D. All of these

26. Mutation was studied in:

A. *Escherichia coli* B. *Neurosphora*

C. Both A & B D. None

27. *In vivo*, growing of tissues in the presence of an antibiotic (*e.g.*, streptomycin, ampilicin, and cycloheximide) is identified using one of the following mutations ?

A. Resistant mutation B. Biochemical mutation

C. Conditional mutation D. Macromutation

28. Mutation frequencies refer to as :

A. Number of mutants present in a population

B. Number of wild type in a population

C. Total plant considered as normal in a polulation

D. Both A & B

29. Physical mutagens include:
   A. Many radiations
   B. Change in pH value
   C. Variable temperature
   D. All of these
30. One nanometer is equal to :
   A. 10^{-9}
   B. 10000000
   C. 10^{10}
   D. None
31. Which rays are known as non-ionizing rays ?
   A. Ultra violet rays
   B. X-rays
   C. Gama rays
   D. All of these
32. Match column A (Radiation source) with Column B (radiation)

Column A	*Column B*
A) Cobalt-60 or Cesium 137	1) gama rays
B) Phosphorus	2) beta rays
C) Nuclear reactor	3) neutrons

   A. A1, B2, C3
   B. B3, A2, C1
   C. B1, A3, C2
   D. None
33. Who discovered mutagenic effect of UV-rays on *Drosophilla* eggs in 1930?
   A. Stadler
   B. Morgan
   C. Altenburg
   D. None
34. Which rays have low penetrating ability?
   A. X rays
   B. Gama rays
   C. Beta rays
   D. UV rays
35. Gama rays can be generated by using of:
   A. Cesium 137
   B. Nuclear reactor
   C. Phosphorus
   D. None
36. EMS mutagens is used to induce mutation in:
   A. Higher plants
   B. Animals
   C. Microorganism
   D. All of these
37. Which of the following character was studied by L. J. Stadler for mutation in maize?
   A. Endosperm character
   B. Leaf characters
   C. Plant height
   D. All of them
38. According to lethal effect of mutation, which progeny will not survive?
   A. Homozygeotes
   B. Heterozygotes
   C. Both A & B
   D. None

39. McClintock described that mutable genes contain transposable elements. This transposable element is known as:

A. Controlling elements
B. Suppressor elements
C. Opron element
D. Operator gene

40. Frame shift mutation occurs when :

A. Addition of bases in DNA
B. Deletion of base in DNA
C. Both A & B
D. None

41. Frame shift mutations cause, which of the following diseases?

A. Tay-Sachs disease
B. Cystic Fibrosis
C. Both A & B
D. Down syndrome's

42. Who developed Singleton's method to detect mutation?

A. Muller
B. W. R. Singleton
C. Both A & B
D. None

43. Which of the following, variety was developed through mutation breeding?

A. PBW 343
B. NW 2036
C. Anupurana
D. Sharbati Sonora

44. Which of the features shows a forward mutation in *Neurospora?*

A. *Ad3* into *ad-3*
B. *His⁻ into his⁺*
C. Both A & B
D. None

45. First generation mutagen treated material is known as:

A. M_1
B. M_0
C. M_3
D. M_2

46. Induced mutation are used for development of:

A. New crop variety
B. Male sterility
C. Double haploids
D. All of these

47. Mutant variety of moong bean developed at Trombay is:

A. TG 37A
B. TPG 41
C. TM 99-37
D. None

48. Todd's Mitcham is a mutant variety of the following crop:

A. Peppermint
B. Barley
C. Both A & B
D. Mung bean

49. BARC is situated at:

A. Mumbai
B. Bangalore
C. Tamil nadu
D. Uttar Pradesh

50. IAEA stands for:

A. International Atomic Energy Agency B. International Atomic equipments Agency

C. National Inter effected Area D. None

51. Who discovered a phenomenon of para-mutation in corn in 1956?

A. Brink B. K. C. Mehta

C. Y. L. Nane D. None

52. Who studied *Ac-Dc* (control element) system in corn chromosome no. 9 in 1950?

A. McClintock B. Stadler

C. Brink D. Kihara

53. McClintock received a Noble Prize in:

A. 1983 B. 1988

C. 2002 D. 1919

54. S. K Luria and M. Delbruck studied bacteriophages and received Noble Prize with A.D. Hershey in 1975 for:

A. Peace B. Medicine

C. Both A & B D. None

55. Based on dominance, mutation are classified into:

A. Dominant Mutation

B. Recessive Mutation and co dominant mutation

C. Incomplete Mutation/partial mutation

D. All of these

56. Auxotrophic mutants are:

A. Do not develop a new biochemical product

B. Capable to develop related compound

C. Unable to synthesize a metabolite

D. Both A & C

57. Transition is a process, by which:

A. Purines are replaced by another purines

B. Pyrimidines are replaced by another purines

C. Purines are replaced by another pyrimidines

D. All of these

58. Transversion refers to the replacement of:

A. Purines by pyrimidines

B. Pyrimidines are replaced by another Pyrimidines

C. Pyrimidines are replaced by another purines

D. Both A & C

59. Which type of protein is produced by frame shift mutation?
A. Non functional protein
B. Functional protein
C. Both A & B
D. None

60. Which types of mutations are more deleterious than base substitution mutation?
A. Nonsense mutations
B. Missense mutation
C. Both A & B
D. Somatic mutation

61. Nonsense mutation occurs due to:
A. Nonsense codon
B. Missense codon
C. Initiation codon
D. None

62. Which codon is responsible for termination of polypeptide chain?
A. Nonsense codon
B. Stop codon
C. Both A & B
D. Initiation codon

63. Mutation breeding include:
A. Physical mutagens
B. Chemical mutagens
C. Both A & B
D. None

64. What is base addition ?
A. Addition of one or more bases in a DNA
B. Release of one or more bases from DNA
C. Both A & B
D. Release only one base from a DNA molecule

65. According to molecular base of mutation, substitution mutations are:
A. Transition
B. Transversion
C. Both A & B
D. None

66. Which of the following mutagens, causes transversion types of mutation?
A. Alkaliting agents
B. Depurination
C. Both A & B
D. None

67. Transition arises due to:
A. Tautomerization and ionization
B. Base analogous
C. Deamination
D. All of these

68. Which, natural base analogous is found in *E. coli* ?
A. 5-hydroxymethyl cytosine
B. 5-glucosyl hydroxymethyl cytosine
C. Both A & B
D. 5-methyl cytosine

69. Prototropy is a condition:
A. most common form of tautomers
B. Relocation of a proton
C. Arises due to change the position of proton
D. All of these

70. Which of the following relationship is obtained between dosage of radiation and number of mutations?

A. Linear B. Colinear
C. Both A & B D. None

71. Which of the following is associated with indirect mutagenic effect?

A. Ultra violet rays B. AP
C. Both A & B D. None

72. Who suggested a technique for detection of biochemical mutations in 1952?

A. J. Lederberg B. Beadle and Tatum
C. Tatum only D. Beadle

73. Woodward' s filtration technique was developed in 1954 by:

A. Woodward B. Tatum
C. Beadle D. None

74. Study of biochemical pathway for the synthesis of eye pigment in *Drosophilla* was made in 1936 by:

A. Beadle and Ephrussi B. Beadle and Tatum
C. Calvin and Hill D. None

75. Which of the following agent, inhibited purine synthesis resulting in mutation?

A. Azaserine B. Urethane
C. Both A & B D. None

76. During replication of DNA, incorporation of AP pairs with of the following :

A. AP-T base pair B. AP-C base pair
C. AP-U base pair D. AP-UU base pair

77. Which of the following factor inhibited pyrimidine synthesis?

A. Urethane B. Nitrogen
C. Both A & B D. None

78. Which of the following is agent considered to alter resting nucleic acid?

A. Nitrous acid (HNO_2) B. NH_2OH
C. Diethyl sulphonate D. All of these

79. Which of the following chemical deaminates Adenine to hypoxanthine, cytosine to uracil and guanine to xanthine?

A. Nitrous acid B. NH_2OH
C. NH_2NH_2 D. None

80. Which of the following agent is responsible for hydroxylation of cytosine to hydroxylcytosine?

A. Nitrous acid B. NH_2OH
C. NH_2NH_2 D. Both A & B

81. AAATTTGGG is converted into GGGTTTGGG as a resultant of mutation is called as:

A. Transition
B. Transversion
C. Tautmerization
D. All of them

82. Hydrazine causes the breaking of uracil rings results in the production of :

A. Pyrazolone
B. 3 aminopyrazole
C. 5-BU
D. Both A & B

83. Which agent causes mutation in bacteria and higher plants?

A. Acridine dyes
B. Acriflavin
C. Allium
D. Both A & B

84. Who proposed target theory of induced mutations in 1935?

A. Timofeeff Ressovsky *et al.*,
B. De Vries
C. Kumar, K.
D. Yaduvada, G

85. A breeder exposes his genetic material (DNA) to low pH (1 to 6), then, which of the following change arises with materials?

A. Base substitution
B. No change occurs at molecular level
C. Backbone breakage of DNA
D. Both A & C

86. Which factor reduces, the frequency of breaking of chromosomes using radiation?

A. Low Oxygen concentration
B. High Oxygen concentration
C. High dose of 2AP
D. None

87. Low oxygen concentration reduces, the frequency of breaking the chromosomes by radiation is called:

A. Anoxia effect
B. Xeroderma effect
C. Photoreactivation
D. None

88. Which condition arises due to wrong pairing of bases in DNA?

A. Transition
B. Transversion
C. Thymine dimer
D. Both A & B

89. Which disorder arises due to change in genetic makeup?

A. Xeroderma pigmentosum
B. Down syndrome
C. Klinnefilter syndrome
D. All of these

90. Which of the following change gives loop formation for chromosomal bases at pachytene sub stage?

A. Deletions
B. Addition
C. Both A & B
D. None

91. How many DNA polymerases were identified till 2004?

A. 2
B. 3
C. 4
D. 5

92. Which protein acts against UV damage budding in yeast (*S. cerevisiae*)?

A. Rad 30 B. umu 30

C. H2a D. All of these

93. Hemoglobin molecule has four polypeptides consisted of:

A. Two identical alpha chain B. Two identical beta chain

C. Both A & B D. Four non identical beta chain

94. Tautomeric thymine pairs with the normal guanine and cytosine with adenine. These pairs are called as:

A. For bidden base pairs B. Unusual base pair

C. Both A & B D. None

95. Genotrophs arises due to:

A. Gene amplifications B. Gene mutation

C. Semi lethal mutation D. All of these

96. Which effect arise on DNA due to exposure of UV rays ?

A. Hydration of pyrimidines B. Formation of thymine dimer

C. Both A & B D. Ionization

97. Who reported occurrence of transversion in 1959 ?

A. Freese B. Morrell

C. Mumun D. None

98. Removal of a purine from the strand of DNA is called:

A. Depurination B. Purinization

C. Stratification D. Scarification

99. Kr is the unit of :

A. beta compound of rays B. gama compound of rays

C. X rays D. None

100. Which of the following rays have high penetrating charge of an organism?

A. Gama rays B. X rays

C. Both A & B D. UV rays

101. How many kinds of mutation are found in *Escherichia coli* ?

A. trp A and trp B B. trp C and trp D

C. trp E D. All of these

102. A process by which base pair of DNA take correct pairing is called as:

A. Synthesis of chromosome B. Replication

C. Semiconservative replication D. All of these

103. Ray has least penetrating capacity?

A. UV rays B. X rays

C. Both A & B D. Gama rays

104. A technique, in which isoalleles are used to assess one of following:

A. Genes contain mutation with small effect B. Phenotypic effect

C. Visual effect D. All of these

105. Mutation may be due to:

A. Recessive alleles B. Dominant alleles

C. Pseudo-alleles D. All of these

106. In cell, recessive mutation will be detectable only in:

A. Homozygous condition B. Heterozygous condition

C. Both A & B D. None

107. Germ cells as a gametic cells carry:

A. Dominant mutation B. Recessive mutation

C. Sub lethal mutation D. All of these

108. Germinal mutation occurs in which of the following stage?

A. Any stage of somatic cell division B. Any stage of reproductive cycle

C. Both A & B D. None

109. Mutation occurs in gametes is called:

A. Germinal mutation B. Gametic mutation

C. Both A & B D. None

110. Who observed dominant germinal mutation in domestic animals?

A. Seth Wright B. T. H. Morgan

C. Swell Wright D. None

111. Which of the following is a classical example of pleiotropy?

A. $Hb^{S}\beta$ B. PKU (phenylketonuria) disease

C. Albinism and sickle-cell anemia D. All of these

112. When a normal or wild type gene converts into mutant gene or recessive types then it is called as :

A. Back mutation B. Forward mutation

C. Dominant mutation D. All of these

113. When a mutant allele converts into normal or wild type allele, then it is called as:

A. Back mutation or reverse mutation B. Forward mutation

C. Dominant mutation D. All of these

114. Who developed antagonist theory of pleiotropy in 1957?

A. G. C. Williams
B. Haeckel
C. Stearns, F. W.
D. Morgan, M.T.

115. Restoration of the original phenotype may occurs due to:

A. Reverse mutation
B. Double mutation is opposite direction
C. Back mutation
D. All of these

116. If the chromosome structure change due to ionization results two breaks occurs. The types of mutation based on which of the theory:

A. Two hit kinetics
B. Single hit kinetics
C. Multiple hit kinetics
D. All of these

117. Protection of the germinal cells from radiation effect is possible only in one of the following:

A. Tissues contain lower oxygen concentration
B. Tissues contain high oxygen concentration
C. Tissues contain no oxygen concentration
D. Radiation non effective by addition of oxygen in tissues

118. Which of the following factor is significantly associated with radiation in change of mutation frequency?

A. Environmental factor
B. Oxygen factor
C. Both A & B
D. None

119. E. S. Robertson and I. C. Anderson described role of environmental factors in induction of mutation in corn plants in:

A. 1961
B. 1970
C. 2009
D. None

Terms

- **Auxotroph Mutation:** Growth can occur by supplementing the medium with specific compound by growth fails and create mutation is known as autotrophic mutation.
- **Back mutation:** A mutant allele converts into normal or wild type allele.
- **Chromosomal mutation:** Mutation occurs within chromosome.
- **Forward Mutation:** The mutation of a wild type gene to mutant phenotypes.
- **Frame shift mutation:** A mutation that upsets the reading frame of mRNA by addition or deletion of nucleotides.

- **Homeosis:** The replacement of one to another caused by mutation or environmental factor.
- **Homeotic mutation:** Mutation caused by hox genes.
- **Hot spot:** A region of chromosome exhibiting high frequency of mutational changes.
- **Induced mutation:** Mutation created by mutagen for various purposes.
- **Isoalleles:** similar or identical alleles.
- **Lethal mutation:** The changes in the genotypes leading to death of an individual.
- **Mutability:** Ability of mutagen to create mutation.
- **Mutable genes:** Genes exhibited high mutation rate.
- **Mutagen:** Environmental, physical and chemical agent capable to create mutation.
- **Mutagenesis:** A process by which cell become mutant.
- **Mutant allele:** An allele represented as recessive and differs from wild type.
- **Mutant site:** A site of organism that altered by mutagen.
- **Mutant:** Product of mutation.
- **Mutation breeding:** Use of mutagens in breeding program to create mutant for improvement of yield of crop plant.
- **Mutation frequency:** The proportion of mutant to the total genotypes of the population.
- **Mutation pressure:** A constant mutation rate that adds mutant gene to a population; due to repeated occurrence of mutations in a population.
- **Mutation:** Organism showed sudden, heritable and discrete variation. In other hand, change in DNA at a particular locus in an organism.
- **Muton:** A part of gene exhibits mutation.
- **Point or gene mutation:** The process by which new alleles of a gene are produced.
- **Protroph mutant:** Cells have capacity of growing at minimal medium is known as protroph mutant.
- **Spontaneous mutation:** Mutation occurs in nature.
- **Tautomeric shift:** Hydrogen atoms can move from one position in a purine or pyrimidine to another position. This process is known as tautomerization.
- **Transition:** The replacement of a purine by one other purine.
- **Transversion:** The replacement of purine by a pyrimidine or vice-versa.
- **Paramutation:** Mutation refers to one allele in a heterozygous condition changes its partner allele permanently.

ANSWERS

Chapter 10: Mutation

1. A	31. A	61. A	91. D
2. A	32. A	62. C	92. A
3. C	33. C	63. C	93. C
4. A	34. D	64. A	94. C
5. A	35. A	65. C	95. A
6. A	36. D	66. C	96. C
7. A	37. A	67. D	97. A
8. A	38. A	68. C	98. A
9. A	39. A	69. D	99. C
10. A	40. C	70. A	100. C
11. D	41. C	71. C	101. D
12. C	42. B	72. A	102. D
13. C	43. D	73. A	103. A
14. A	44. A	74. A	104. D
15. A	45. A	75. A	105. D
16. A	46. D	76. A	106. A
17. A	47. C	77. A	107. A
18. C	48. A	78. D	108. B
19. A	49. A	79. A	109. C
20. B	50. A	80. B	110. A
21. C	51. A	81. A	111. D
22. D	52. A	82. A	112. B
23. A	53. A	83. D	113. A
24. A	54. B	84. A	114. A
25. D	55. D	85. D	115. D
26. C	56. D	86. A	116. A
27. A	57. A	87. A	117. A
28. A	58. D	88. D	118. C
29. D	59. A	89. D	119. A
30. A	60. A	90. C	

Chapter **11**

Gene (DNA, RNA and Alleles)

1. Who concluded that pepsin could not break cell nucleus ?
   A. F. Miescher B. A. Kossel
   C. Levene D. None
2. According to Miescher, which molecule does not contain sulphur ?
   A. Nuclein B. DNA
   C. Both A & B D. None
3. Who awarded Noble Prize in 1970 for demonstration of bases in nucleic acid?
   A. A. Kossel B. P. A. Levene
   C. Griffith D. None
4. Who demonstrated sugar molecule in nucleic acid?
   A. A. Kossel B. P. A. T. Levene
   C. Griffith D. Tatum
5. How many stains taken in experiment of Griffith in 1928?
   A. 2 B. 3
   C. 4 D. 6
6. A way by which transferring a trait of a stain to any other through using a DNA take out of the former is called:
   A. Transduction B. Special transduction
   C. Conjugation D. None
7. DNA does not contain:
   A. Sulphur molecule B. Nitrogeneous bases
   C. Phosphorus compound D. None

8. Tobacco Mosaic Viruses does not contain:
   A. DNA B. RNA
   C. Both A & B D. None
9. Which of the material is used by H. Frankel-Conrat for identification of RNA is a genetic material?
   A. Tobacco Mosaic Virus B. Bacteriophage (T2)
   C. Both A & B D. None
10. A way by which transferring a trait of a stain to any other using a DNA is taken out of the former called:
   A. Transduction B. Special transduction
   C. Conjugation D. Transformation
11. Part of bacteriophage is:
   A. Tail B. Head
   C. Both A & B D. Tail fibres
12. In bacteriophase, tail is consisted of:
   A. Tail core B. Tail sheath
   C. Tail fibres D. All of these
13. Who experimented that the DNA is the sole genetic material in DNA bectriophage?
   A. Grifith B. Linus Pauling
   C. Hershey and Chase D. None
14. Nucleic acid is known as DNA as well as:
   A. RNA B. Protein
   C. Amino actids D. None
15. Who introduced the term of prion in 1982 and also received Noble prize in 1977?
   A. Stanley B Prusiner B. Jece Lee
   C. S. E. Luria D. Delbruck
16. Which is related to prion?
   A. Traits as proteinaceous infected particles
   B. Generally fatal impression
   C. Causing various neurodegenerative disease
   D. All of these
17. Mad cow disease is caused by:
   A. Prion B. Bacteria
   C. Fungus D. None
18. Symbal of prion protein is:
   A. PrP B. PPP
   C. PPPr D. AUG

19. BSE stands for:
    A. Bovine spongiform encephalopathy B. BATA sau ellopee
    C. Both A & B D. None
20. Which of the following showed protein conformation base inheritance?
    A. Prion B. Virusoids
    C. Retrovirus D. None of these
21. Who experimented on protein based inheritance in infected particles and also received Noble prize in 1977?
    A. Stanley B Prusiner B. Jece Lee
    C. S. E. Luria D. Delbruck
22. Viruses have their own:
    A. DNA/RNA B. Proteins
    C. Both A & B D. None
23. Molecular weight of protease resistant core of PrP is:
    A. 27-30kD B. 10kD
    C. 5kD D. None of these
24. PrP^{C} is a subunit of PrP^{Sc}, is rich in :
    A. α- helices B. β-sheets
    C. Both A & B D. None
25. Based on sugar, RNA have:
    A. Cytidine B. Uridine
    C. Guanosine and Adenosine D. All of these
26. A polymeric chain composed by many monomer unit is called:
    A. PCR B. Nucleoside
    C. Nucleotide D. All of these
27. Jointing of nucleoside with one phosphate molecule resulting the formation of:
    A. Nucleotide B. Nucleic acid
    C. DNA D. RNA
28. In formation of nucleotide, phosphate molecules is joint with sugar at:
    A. Carbon no. 5' or 3' B. Carbon no. 1'
    C. Carbon no. 2' D. All of these
29. Bond found between two nucleotide:
    A. Phosphodiester bond B. Glycolic bond
    C. Nonsense bond D. None
30. Who suggested DNA is found in helical form ?
    A. Linus Pauling B. Watson
    C. Crick D. All of these

31. Who indicated as A = T and G = C ?

A. Linus Pauling
B. Watson
C. Crick
D. E. Chargaff

32. In DNA, A = T and G = C is based on :

A. Law of Inheritance
B. Chargaff rules
C. Linus rules
D. None

33. Which form of DNA have right handed turns?

A. B-DNA
B. Z-DNA
C. Both A & B
D. None

34. Which form of DNA have left turns?

A. B-DNA
B. Z-DNA
C. Both A & B
D. None

35. A locatable region of genomic sequence, corresponding to a unit of inheritance, which is associated with regulatory regions, transcribed regions, and or other functional sequence regions is known as :

A. Gene
B. Factor
C. Traits
D. All of these

36. One kb is equaled to:

A. 1000 base pairs
B. 100000base pairs
C. 50000 base pairs
D. 10 base pairs

37. A kilo base (kb) is a unit of measurement in molecular biology equal to :

A. 1000 base pairs of DNA or RNA
B. 200 base pairs of DNA or RNA
C. 300 base pairs of DNA or RNA
D. 10 base pairs of DNA or RNA

38. One complete helix of Z-DNA is equal to:

A. 45 Å
B. 40 Å
C. 90 Å
D. None

39. Accordingly, RL model, which of the following sentence is correct?

A. B-DNA is not always right handed
B. B-DNA is always right handed
C. B-DNA is also be left handed
D. Both A & C

40. Which sentence is true for Z-DNA?

A. DNA follows Zig-zag form
B. It is left handed
C. Residues per turn is 12
D. All of these

41. Which is not true?

A. B-DNA have 10 residue per turn
B. Z-DNA have 12 residue per turn
C. Diameter of B-DNA is 18 Å
D. Diameter of Z-DNA is 18 Å

42. B-DNA have :

A. Right handed

B. 3.4 Å rise per residue and rotation per residue is 36°

C. Diameter of B-DNA is 20 Å and 6° base per tilt

D. All of these

43. Z-DNA have which of the following ?

A. Left handed helical and rotation per residue is 60°

B. 12 residue per turn and 18 Å in diameter

C. 45 Å helix per pitch and 7° base per tilt

D. All of these

44. Based on helix , how many form of DNA is found?

A. A | B. C

C. B and Z | D. All of these

45. A and B form of DNA is obtained from:

A. Z-DNA | B. B-DNA

C. D-DNA | D. All of these

46. Positive super coils represents:

A. Over-winding (superhelical) | B. Can be created in vitro

C. Both A & B | D. None

47. Super coiling controlled by:

A. Linking number | B. Twisting number

C. Writhing number | D. All of these

48. Negative super-coils directions may be:

A. Opposite to the clockwise turn of B-DNA

B. To clockwise turn of B-DNA

C. Both A & B

D. None

49. Nature of RNA is:

A. Double stranded | B. Single stranded

C. Both A & B | D. None

50. Which of the following sentence is not true regarding virusoids?

A. Not replicated self | B. Self replicated organism

C. Protein coated organism | D. All of these

51. Which is true?

A. mRNA = 5% to 10% of total RNA | B. tRNA = 10% to 15 % of total RNA

C. rRNA = 80% of total RNA | D. All of these

52. Small nuclear RNA is found in:
A. Nucleus
B. Cytoplasm
C. ScRNA
D. None

53. In 1953, Duplex DNA was proposed by:
A. Watson
B. Crick
C. Both A & B
D. Linus Pauling

54. Who observed the possibility of triplex DNA?
A. Linus Pauling
B. Levene
C. Strasburger
D. A. Rich

55. A process by which DNA directed DNA synthesis:
A. DNA replication
B. Reverse transcription
C. Both A & B
D. Transformation

56. All cells of a living organism have:
A. Same quantity and quality of DNA
B. Different quantity of DNA and quality same
C. No same quality of DNA
D. None

57. Replication of double stranded DNA is done by:
A. Conservative
B. Semi-conservative
C. Dispersive
D. All of these

58. Which fact is true for conservative DNA replication?
A. dsDNA molecule is conserved and new molecule is synthesized from old molecule
B. Old molecules is disintegrated and two new molecules can be synthesized
C. Both A & B
D. None

59. Which fact is true for conservative DNA replication?
A. dsDNA molecule is conserve and new molecule is synthesized
B. Old molecules is disintegrated and two new molecules can be synthesized
C. Both A & B
D. None

60. Who demonstrated semi-conservative mode of replication of bacterial chromosome (prokaryotes)?
A. J. Cairns
B. F. W. Stahl
C. M. Meselson
D. None

61. Who developed experiment on autoradiography in Vicia faba which published in 1957?
A. J. H Taylor and Co-worker
B. Cairns
C. Morgan
D. Stadler

62. Harlequin chromosome is usually suitable for the study of:

A. Semi-conservative replication B. Crossing –over

C. Sister chromatids exchanges D. All of these

63. Bidirectional DNA replication is also found in:

A. Yeast B. Fruitfly

C. Mammals D. All of these

64. Why, DNA polymeraze is not initiated the synthesis the DNA?

A. Process required primer strand

B. Primer is breakdown the capacity of DNA polymerase, results not capable for DNA synthesis

C. Not required of primer

D. None of these

65. RNA is synthesed from template DNA with the help of :

A. Primer B. Protein

C. DNA polymerase D. None

66. How many families of DNA polymerase based on phylogenetic relationships?

A. 6 B. 4

C. 2 D. 10

67. How many prokaryotic DNA polymerase ?

A. Five B. Six

C. Seven D. None

68. Who discovered DNA pol I in 1956?

A. Arthur Kornberg B. N Kornberg

C. Khurana D. Jones

69. How many active sites have DNA polymerases I?

A. Five B. Six

C. Seven D. None

70. Which enzyme mainly involved in removing RNA primers from Okazaki fregments and fills up the gaps due to its 5 prime to 3 prime polymerizing capacity?

A. DNA pol I B. DNA pol III

C. Both A & B D. None

71. Which enzyme is used for covalent bonding after the gaps filled by DNA pol I?

A. Helicase B. Ligase

C. Polymerase D. All of these

72. DNA polymerase II is associated with :

A. B family of DNA polymerase B. A family of DNA polymerase

C. Both A & B D. D family of DNA polymerase

73. Which enzyme is resembles to DNA polymerase I ?

A. DNA polymerase II
B. DNA polymerase III
C. DNA polymerase IV
D. All of these

74. Which enzyme belongs to family of C of DNA polymerase and play a essential role in DNA replication ?

A. DNA pol III
B. DNA pol I
C. DNA pol II
D. All of these

75. Who formulated replicon model of DNA replication in 1963?

A. Brenner and Cuzin
B. Kornberg
C. Chase
D. Jacob

76. Number of replicons found in E. colis :

A. One
B. Three
C. Four
D. None

77. How many replicons present in yeast:

A. 500
B. 1000
C. 10000
D. 20

78. Site of replication is observed in:

A. Bacteria
B. Yeast
C. Chloroplast and mitochondria
D. All of these

79. Origin of replication is rich in:

A. A and T
B. G and T
C. U and G
D. None of these

80. Ori C is a region of replication present on:

A. E. coli
B. Diplococuss spp.
C. Both A & B
D. None

81. How many base pairs present in Ori C of E.coli:

A. 245
B. 289
C. 500
D. None of these

82. Who dêmonstrated the semi-conservative model of DNA replication in 1958?

A. Meselson and Stahl
B. Crick and Watson
C. Cairns, J
D. None of these

83. Who observed replication fork during 1963?

A. J. Cairns
B. Crick
C. Watson
D. None

84. Which is true regarding replication fork?

A. One DNA molecule appears to be branched into two molecule

B. Formed Due to progressive separation of the two strands of repeating DNA molecule

C. Both A & B

D. None

85. Replicon is:

A. A DNA molecule

B. Capable of DNA replication

C. Terminus as specific site

D. All of these

86. How many origin of replication present in phage T_7 ?

A. Two

B. Six

C. Seven

D. None

87. Which enzyme is used in melting process of DNA?

A. DNA helicase

B. DNA gyrase

C. Both A & B

D. None

88. A process by which DNA molecule show unwinding of two complementary strands:

A. Melting

B. Replication

C. Reverse transcription

D. All of these

89. Which is true in respect of primase?

A. Initiate transcription

B. Terminated transcription

C. Elongation of translation

D. All of these

90. In prokaryotes, replication fork is associated with:

A. Plasma lemma

B. Nuclear envelope

C. Both A & B

D. Ribose

91. In eukaryotes, replication fork may be associated with:

A. Plasma lemma

B. Lysosome

C. Peroxisome

D. Nuclear envelope

92. Which direction of DNA show discontinuous replication?

A. 5 prime to 3 prime

B. 3 prime to 5 prime

C. Both A & B

D. None

93. Which is true in respect of continuous DNA replication?

A. Its directed 5 prime to 3 prime

B. Its directed 3 prime to 5 prime

C. Both A & B

D. None

94. Seq A protein is bind to :

A. Hemi-methylated DNA

B. DNA gyrase

C. Prepriming complex

D. None of these

95. Synthesis of RNA primers by DnaG primase take place and lead to formation of :
A. Prepiming complex
B. Primer complex
C. Holoenzyme
D. Core enzyme

96. Which enzyme is large multiple protein complex elongate of DNA replication in E. Coli ?
A. DnaG
B. DnaB
C. DNA polymerase HE
D. None of these

97. Which is major helicase ?
A. Dna G
B. DnaB
C. Both A & B
D. None

98. Which fact is true in respect of DNA helicase?
A. Unwinding the DNA
B. Translocate of DNN
C. Its also worked as similar to motor protein
D. All of these

99. DNA polymerase HE consisted of which of the following?
A. Polymerase (Core ($\alpha\varepsilon\theta_2$)
B. A sliding clamp (β_2)
C. A clamp loader or match marker (γ Complex)
D. All of these

100. Protein involved in DNA elongation:
A. DNA Gyrase (a topoisomerase)
B. SSB and DNA ligase
C. RNaseH and DNA pol I
D. All of these

101. Use of DNA pol I in replication:
A. Use for unwinding of DNA helix
B. Filled the gap created due to removal of RNA primers
C. Both A & B
D. None

102. DNA ligase is:
A. Convert Okazaki fragment into leading strand
B. Join the two point of break DNA
C. Both A & B
D. None

103. Which enzyme is used for removing of RNA pimers during replication of DNA?
A. RNase H
B. DnaB
C. DnaA
D. DNA pol I

104. DNA primase is associated with DnaB and synthesis of multiple primers for:
A. Lagging strand
B. One primer for leading strand
C. Both A & B
D. None

105. Event came under replication of DNA showed more complexity :
A. Leading strand
B. Lagging strands
C. Both A & B
D. None

106. How many terminator site on E. coli chromosome?
A. Six
B. Seven
C. Eight
D. Hundreds

107. Who proposed the model of semi-conservative DNA replication?
A. Crick and Watson
B. Messelson and Stahl
C. Mose
D. Jensen

108. DNA replication may not be completed without:
A. DNA repair
B. Linearity
C. Both A & B
D. None

109. In eukaryotes, DNA replication show that:
A. Nucleus is the site of replication and S-phase time
B. More complex mechanism
C. Bidirectional
D. All of these

110. In prokaryotes, which fact is true regarding DNA replication:
A. Bidirectional
B. Occur in cytoplasm
C. DNA pol I, II and III enzyme involved
D. All of these

111. Gap found between two RNA pimers is called:
A. Okazaki fragment
B. RL segment
C. Initiation site or origin
D. Terminator site

112. A special feature of ribonucleoprotein molecule use a special mechanism for the synthesis of DNA:
A. Telomerase
B. Tripsin
C. Nob
D. SAT

113. Which strand synthesize telomerase?
A. G-rich strand of Telomers
B. A-rich strand of telomers
C. C-rich strand of telomers
D. All of these

114. Actinomycin D prevents the
A. Synthesis of RNA from DNA
B. Synthesis of RNA from RNA template
C. Both A & B
D. None

115. Which process is responsible to changes of DNA composition?
   - A. Single base changes *i.e.*, deamination of 5methylcytosine
   - B. Structural distortions
   - C. Both A & B
   - D. None

116. RNA synthesis is stopped by the use of :
   - A. Rifampicin
   - B. Chloramphenicol
   - C. Both A & B
   - D. None

117. Which of the following drugs is inhibit the action of Rifampicin?
   - A. Chloramphenicol
   - B. Rifampicin
   - C. Both A & B
   - D. None

118. Polymerases of E.coli are designed by:
   - A. Roman numbers
   - B. Counting numbers
   - C. Tables
   - D. None

119. Polymerases of eukaryotes are designed by:
   - A. Greek letters
   - B. Latin letter
   - C. British letter
   - D. German letter

120. DNA replicase is also known as:
   - A. DNA polymerase
   - B. Dna G
   - C. Both A & B
   - D. None

121. DNA polymerase II and III discovered in:
   - A. Pol A mutant of E. coli.
   - B. Neurospora
   - C. Oenothera
   - D. None

122. Pol A mutant of E. coli is deficient of:
   - A. DNA Polymerase I
   - B. DNA polymerase II
   - C. DNA polymerase III
   - D. All of these

123. DNA polymerase I has unique property in which of the following?
   - A. Encodes gene pol A
   - B. Nick translation
   - C. Both A & B
   - D. None

124. Enzyme degrades the nucleic acid form one end is called:
   - A. Endonuclease
   - B. Exonuclease
   - C. Pepsin
   - D. None

125. Primase is:
   - A. RNA polymerases
   - B. Catalyzes the synthesis of RNA primers
   - C. DnaG proteins
   - D. All of these

126. Restriction endonuclease play an important role in:

A. DNA replication
B. DNA repair
C. Both A & B
D. None

127. Who developed the fine structure of gene map in 1955?

A. Benzer
B. Bateson
C. Muller
D. None

128. Accordingly Benzer, one of the following parts of gene is:

A. Cistron
B. Muton
C. Recon
D. All of these

129. A sequence of nucleotides arranged in a linear/ring fashion on the DNA polymer is called as:

A. Gene
B. Traits
C. DNAs
D. None

130. Seymour Benzer prepared a fine structure of gene working on:

A. Virus
B. Bacteriophage T4
C. Fungus
D. Prion

131. Cistron is a unit of gene as well as:

A. Unit of functional
B. Unit of replication
C. Unit of mutation
D. None

132. Muton is a part of gene responsible for:

A. Mutation
B. Replication
C. Transcription
D. Chromosome synthesis

133. Recon is unit of:

A. Recombination
B. Mutation
C. Protein
D. None

134. Gene complementation test can be used for:

A. To study function of gene
B. To study of structural complexity of gene
C. Both A & B
D. None

135. A segment of DNA in the chromosome which carries the information for directing the synthesis of a single polynucleotide chain is known as:

A. Cistron
B. Recon
C. Both A & B
D. Muton

136. Which of the following segment of DNA is small found between two mutation sites?

A. Recon
B. Cistron
C. Muton
D. None

137. In eukaryotes, which of the following sequence is not translated?
A. Coding sequences
B. Intervening sequences
C. Both A & B
D. None

138. In eukaryotes, which one of the following is the product of primary transcripts?
A. hnRNA
B. mRNA
C. Protein
D. All of these

139. Length of hnRNA varies:
A. 500-1000 nucleotides
B. 10-20 nucleotides
C. 200-50 nucleotides
D. None

140. The α gene family is located on one of the following chromosomes of human being:
A. Chromosomes no 16
B. Chromosomes no. 46
C. Both A & B
D. None

141. All the globin genes similar in respect of :
A. Structural
B. Have three exons
C. Have two intron
D. All of these

142. Antibody molecule is made up of :
A. Four immunoglobulin
B. Two immunoglobulin
C. Three immunoglobulin
D. None

143. Who accomplished chemical synthesis of gene in 1970?
A. Agrawal and Khurana
B. Mc Clintock
C. K Harvindra
D. None

144. Which part of gene responsible to synthesis of a single protein?
A. Cistron
B. Muton
C. Recon
D. None

145. Number of introns lies between:
A. 2-7
B. 10-15
C. 500-800
D. None

146. How may introns found in globin gene?
A. 2
B. 3
C. 5
D. 7

147. How many introns found in chick albumin gene?
A. 7
B. 2
C. 90
D. 12

148. How many introns found in alfa collagen gene?
A. 50
B. 2
C. 7
D. None

149. Which of the DNA has maximum genetic information?

A. Unique DNA
B. Repetitive DNA
C. Both A & B
D. None

150. A haploid cell/basic chromosome of an individual has total amount of DNA is known as:

A. GH vale
B. C-value
C. F test
D. None

151. The total number of DNA nucleotide residues in the haploid genome is known as:

A. D DNA
B. C DNA
C. C value
D. None

152. Hyperchromicity occurs in DNA when the:

A. DNA duplex is denatured
B. DNA duplex is normal
C. Newly synthesis DNA
D. None

153. SINES stands for:

A. Short interspersed sequences
B. Short intron sensors
C. Short newly sequences
D. None

154. Repetitive DNA consists of:

A. SINES
B. LINES
C. Retroposons
D. All of these

155. Density of DNA of Bacteria is one of the following:

A. 10 mg ml^{-1}
B. 100 mg ml^{-1}
C. 500 mg ml^{-1}
D. None

156. Density of DNA is found in eukaryotic nucleus:

A. 10 mg ml^{-1}
B. 100 mg ml^{-1}
C. 500 mg ml^{-1}
D. None

157. Stable proteins in the chromosome are:

A. H1 and H4
B. H2A
C. H2B and H3
D. All of these

158. Histones H3 and H4 composed together in form of dimer to have a size equal to:

A. HI
B. H2A
C. H2B
D. None

159. Core particles present in:

A. H2B and H2A
B. H3 and H4
C. HI
D. All of these

160. The length of DNA varies from:

A. 154 bp to 260 bp
B. 10 bp to 2o bp
C. More than 4000 bp
D. None

161. A technique used to find out the distances of histones and DNA form centre of nucleosome is known as:

A. Neutron scattering technique
B. Photon scatter technique
C. Both A and B
D. None

162. DNA is wrapped by:

A. Histones protein
B. Chromosomes
C. Matrix
D. All of these

163. Enzymes used for making single strand nicks in DNA are:

A. DNase I
B. DNase II
C. Both A & B
D. None

164. SV 40 have a circular molecule of DNA with length as:

A. 5200 bp long or 1500 nm long
B. 20 bp long
C. 100 bp long
D. None

165. Linking number of paradox is :

A. Number of times two DNA strands are interwined in circular DNA
B. Structures of the nucleosome displayed an over-twisted left-handed wrap of DNA around the histone octamer
C. Both A & B
D. None

166. Linker histone H1 is responsible for:

A. Proper folding of chromatin fibres by three domain
B. Unwinding of DNA
C. De-naturation of DNA superhelix
D. All of these

167. Nucleosome is the substrate of :

A. DNA replication
B. DNA recombination
C. Transcription and repair of DNA
D. All of these

168. Both acetylation and methylation are helpful for assembly of nucleosome in:

A. S phase
B. G1 phase
C. Metaphase
D. Anaphase

169. Hyperacetylation is:

A. Increase the accessibility of nucleosomal DNA to transcription factors
B. Decrease the accessibility of nucleosomal DNA to transcription factors
C. Increase the transcriptional silencing
D. None

170. Which of the following enzyme play a key role in regulation of transcription?

A. Histrone acetyltransferases B. Deacetylases

C. Both A & B D. None

171. Who discovered spilt gene in 1977?

A. Phillip A. Sharp and Richard J. Roberts

B. Barrell

C. Nirenberg MW

D. None

172. Who won the Noble Prize in Medicine or physiology of split gene in 1993?

A. Phillip A. Sharp and Richard J. Roberts B. Barrell

C. Nirenberg MW D. None

173. How many amino acid consisted by ovalbumin gene?

A. 387 B. 386

C. 600 D. 7000

174. Who used the term exon and intron?

A. W Gilbert B. M Mamoina

C. Dillonie F D. None

175. Interrupted genes include:

A. Nuclear gene for protein and nuclear genes coding rRNA or tRNA

B. Mt Gene in Yeast

C. Chloroplast genes in plant

D. All of these

176. How many exon and introns found in split gene of ovalbumin ?

A. 8 exons and seven introns B. 23 exon and 1 introns

C. Only 19 exons D. None

177. Plant introns are differs from plant exons as:

A. Plant introns are richer in AT base pairs

B. Plant exons are richer in AT base pairs

C. Plant introns are resamble to fruit fly

D. None

178. Who developed early intron hypothesis in 1987?

A. Walter Gilbert B. Barrell

C. Chase and Hurshey D. Mc Charty

179. Early intron hypothesis is also known as:

A. Exon theory of gene B. Barrel theory of gene

C. Both A & B D. None

180. Late intron hypothsis is also known as:
A. Insertional theory of intron origin B. Exon theory of gene
C. Barrel theory D. None

181. The genome of xX174 consist of :
A. Nine cistrons B. 19 cistrons
C. 20 cistrons D. 50 cistrons

182. Who observed nine cistrons in xX174 in 1976?
A. Barrell B. W. Gilbert
C. Sanger, F D. Crick

183. Which of the following overlapping genes is reported in human being?
A. MUTYH B. TOE 1
C. TESK 2 D. All of these

184. A DNA sequence that resembles a homologous normal gene:
A. Dominant gene B. Pseudogene
C. Recessive genes D. Wild type

185. Which gene does not cause any disease or harmful effect on an organism?
A. Pseudogenes B. Dominant gene
C. Recessive gene D. None

186. Retropseudogenes created by:
A. Reverse transcription B. Translation
C. Replication D. Melting of DNA super helix

187. Retropseudogenes lack of :
A. Introns B. Exons
C. Both A & B D. None

188. Examples of retropseudogenes in human being are:
A. SINES B. LINES
C. Both A & B D. None

189. Retrogenes originate due to:
A. Retrotransposition B. Reverse transcription
C. Both A & B D. None

190. Retrogenes have been identified in:
A. *Drosophilla* B. Human
C. Mouse D. All of these

191. A phenotypically silent DNA sequences that are not expressed during the formal life cycle of an organism is known as:
A. Cryptic gene B. Selfish gene
C. Pseudogenes D. None

192. Cryptic gene activated in population due to:

A. Mutation
B. Transposition
C. Recombination
D. All of these

193. DNA differs from RNA in:

A. Bases
B. Sugar and bases
C. Sugar and phosphate
D. Sugar, bases and phosphate

194. The distance between two base pairs in B-DNA is:

A. 34 nm
B. 3.4 nm
C. 0.34 nm
D. None

195. Maximum recombination frequency between two gene is:

A. 50%
B. 100%
C. 75%
D. 20%

196. *RecA* is a protein involved:

A. Recombinational repair system
B. Mismatch repair
C. Nucleotide excision repair
D. Base excision repair

197. A gene is said to be dominant if:

A. It expresses its effect only in homozygous condition
B. It expresses its effect only in heterozygous condition
C. Both A & B
D. Never expresses its effect in any condition

198. Gene which code for a pair of contrasting traits are known as:

A. Factor
B. Gamete
C. Allele
D. None

199. A number of alternative phases of a gene is known as:

A. Allele
B. Factor
C. DNA
D. All of these

200. How much phases encoded one gene?

A. Two
B. Three
C. Four
D. Six

201. Who discovered ABO blood group in human being in 1900?

A. Landsteiner, K.
B. Johansson
C. Jensen
D. Baron Emil von Dungern

202. Who proposed the heritability of ABO blood group in 1910-11?

A. Ludwik Hirszfeld and E. von Dungern
B. Felix Bernstein
C. Jan Janský
D. K. Landsteiner

203. Who demonstrated the correct blood group inheritance pattern of multiple alleles at one locus in 1924?

A. Felix Bernstein
B. Morgan
C. Boveri and Sutton
D. McClintock

204. Examples of multiple alleles is:

A. ABO blood group in human being
B. Coat color in rabbit
C. Eye color in Drosophilla
D. All of these

205. Which test is to be used for detection of alleles?

A. Recombination test
B. Cis-trans test
C. Both A & B
D. None

206. What is not sub divisible?

A. Gene
B. Factor
C. Character
D. None

207. Crossing over do not occurs in:

A. Gene
B. Male Drosophilla
C. Silk moth
D. All of these

208. Who coined the term multiple alleles in 1914?

A. T. H. Morgan
B. Landsteiner
C. Johansson
D. Mandel

209. Which sign is represented wild type alleles?

A. Positive sign
B. Negative sign
C. Circle sign
D. None

210. Who described the self sterility in *Nicotiana sanderae* in 1925?

A. East and Mangeldorf
B. Race and Sanger
C. Wiener
D. Morgan

211. Who discovered Rh blood group antigen system in 1940

A. Landsteiner and Wiener
B. Landsteiner and Morgan
C. East and Shull
D. Khurana H and Crick

212. Snapdragon is a good example of multiple allele in plant observed by:

A. Baur
B. Race and Sanger
C. Wiener
D. Morgan

213. Gene for albinism is:

A. Recessive
B. Dominant
C. Co-dominant
D. Partial dominant

214. A viable DNA (deoxyribonucleic acid) coding that occupies a given locus (position) on a chromosome is known as:

A. Allele B. Gene factor

C. DNA D. Factor

215. Who demonstrated the intragenic recombination in lozenge gene of Drosophilla in 1940?

A. Oliver B. Baur

C. Both A & B D. Green and Green

216. Strongly coupled and functionally related gene is called as:

A. Pseudoalleles B. True allele

C. Wild allele D. None

217. Nearby more alleles at single locus is called as:

A. Multiple alleles B. Pseudo allele

C. Isoallele D. Normal allele

218. DNA ligase is encoded by which of the following?

A. *E.coli* B. *Phage* T4

C. Both A and B D. None

219. Which of the following optimum time is required for ligation of nicked DNA?

A. 37° C B. 60° C

C. 40° C D. None

220. Bromouracil induces the mutations because it:

A. Replaces a T and bind to G B. Replaces a G and bind to A rather than C

C. Changes the binding affinity of G D. Changes the binding affinity of T

221. Who observed the replication fork in *Escherichia coli* by using the technique of autography in 1963?

A. J. Cairns B. Crick

C. Meselson and Stalh D. None

222. Replication form (RF) occur in which process:

A. Rolling circle replication B. Bidirectional DNA replication

C. Replication fork formation D. None

223. Which is true in respect of rolling circle replication of DNA?

A. It occurs viruses having single stranded DNA

B. Occur during conjugation

C. A single repeating unit from the genome is converted into rolling circle

D. None

224. Which function is associated with DnaA protein?
   A. Origin recognition and melting, initiate the process of prè printing
   B. Fork movement by duplex unwindimg
   C. Primer synthesis , activated by DnaB
   D. Stabilizes single strands by preventing duplex formation

225. Primrose is consisted which of the following protein need to be activated by DnaB for priming:
   A. DnaG   B. DnaB
   C. DnaC   D. DnaT

226. Which of the following protein is consisted of primrose?
   A. Pri A, Pri B and Pri C   B. DnaT and DnaB
   C. DnaC   D. All of these

227. Which protein is functionally as helicase?
   A. DnaB   B. DnaG
   C. DnaC   D. DnaT

228. DNA polymerase is as:
   A. DNA replicase   B. Replicon
   C. DNA replication inhibitor   D. None

229. Which of the enzyme is used for addition of nucleotides to a pre-existing primer in DNA repiication?
   A. DNA polymerase   B. Primase
   C. RNA polymerase   D. None

230. Which of the following enzyme docs not play a role in initiation of synthesis of a polynucleotide in repplication?
   A. DNA polymerase   B. RNA polymerase
   C. Helicase   D. Primase

231. Which of the subunit of polymerase has facilitate the proof reading activity?
   A. ε subunit   B. α subunit
   C. Both A & B   D. None

232. The core enzyme consisted of which of the following unit?
   A. α B. ε
   C. θ D. All of these

233. Which of the following radiation is responsible to formation of abnormal bond in DNA?
   A. UV light   B. X rays
   C. Beta Rays   D. None

234. Which of the following is not a part of nucleic acid?

A. Nucleotide B. Ribose
C. Phosphoric acid D. Polypeptide

235. RNA-prime RNA synthesis as direct replication of RNA occur in:

A. Polio virus B. dsRNA phages
C. Both A & B D. None

236. Ribosomes similar to those of bacteria found in:

A. Plant nuclei B. Cardiac muscle cytoplasm
C. Liver endoplasmic reticulum D. Neuronal cytoplasm

237. Which of the following unit is known as basic unit of nucleic acid?

A. Nucleotide B. Gene
C. Phosphoric acid D. All of these

238. The term nucleic acid proposed in 1889 by :

A. Meischer B. Altman
C. Jeffery D. None

239. Who found nuclein in cell nucleus in 1871?

A. F. Meischer B. Altman
C. Holiker D. R. Brown

240. Which of the following molecule is not presented in nucleic acid?

A. Phosphoric acid B. Sugar
C. Nitrogenous base D. All of these

241. Monomeric unit of nucleic acid is called as:

A. Nucleotide B. Nucleoside
C. Nitrogenous base D. None

242. Which of the following type of nitrogenous base found in nucleic acid?

A. Pyrimidines B. Purine
C. Both A & B D. Leucine

243. How much main nitrogeneous bases madeup nucleic acid?

A. Three B. Four
C. Two D. None

244. In which, thymine is replaced by uracil?

A. RNA B. DNA
C. Amino acid D. Genetic code

Terms

- **DNA:** Deoxyribonucleic acid that carrying the information in several generations of an individual.
- **Additive allelic effect:** Genetic factor that raise/lower the value of a phenotype on a linear scale of measurement.
- **Dosage compensation:** A phenomenon in which the activity of a gene either increases or decreases according the situation.
- **Gene conversion:** A process in which one allele is replicated at the expense another leading to non Mendelian segregation ratios.
- **Gene:** A determinant of hereditary or a unit of inheritance.
- **Genetics:** A branch of biological science deals heredity and its variation.
- **Multigene family:** A cluster of gene those are similar in nucleotide sequence.
- **Multiple allele:** A condition in which a particular gene occurs in three or more allelic form in a population of organism.
- **DNA backbone:** It is a polymer with an alternating sugar-phosphate sequence. In other words, a structure made by sugar and phosphate linkup in alternate form.
- **Nuclear DNA:** The DNA found in nucleus.
- **Circular DNA:** This type of DNA found in prokaryotes. It is ringing shaped DNA.
- **Transformation:** A process of recombinations of the genetic materials as naked DNA form one cell enter and integrated in another cell.
- **Nucleic acid:** Acid found in nucleus *i.e.*, DNA and RNA.
- **Genetic materials:** The materials found in cell contain genetic information/code of an organism.
- **RNA:** Ribonucleic acid as genetic materials in viruses or prokaryotes.
- **Bacteriophage:** A phage is viruses which attack the bacteria and kill them. It has two part *i.e.*, tail and head.
- **Transduction:** The genetic recombination in bacteria in which DNA is transferred from one bacterium to another *via* the bacteriophage. It is of two type (*i*) general transduction (*ii*) specialized transduction.
- **General transduction:** Any part of donor DNA transferred to the recipient DNA.
- **Gene cloning:** Insertion of a fragment of DNA, containing of gene, into a cloning vector and subsequent propagation of a recombinant DNA molecule in a host organism.
- **Gene library:** A large collection of cloning vectors containing a complete set of fragments of the genome of an organism.
- **Gene sequencing:** The technique of determining the order of bases of DNA molecule which constitute a gene.
- **Gene silencing:** A process in which an endogenous gene is expressed by introduction of a related transgene.

- **Gene synthesis:** Chemical synthesis of a gene involving blocking of certain reactive groups of nucleotides and joining the same one by one.
- **Gyrase:** An enzyme that relax the super coiling caused by unwinding of double helix of DNA.
- **Hairpin loop:** A double helical region formed by base pairing between adjacent complementary sequence in a single strand of DNA or RNA.
- **Hapten:** A small molecule that acts as an antigen when conjugated to a protein.
- **Hetero allele:** The allele which arise due to base pair replacement in different mutons.
- **Helicase:** An enzyme that acts a key role during DNA replication for breaking the hydrogen bonds those hold the double helix together.
- **Heterodimer:** A dimeric protein having two different polypeptides.
- **Histone:** One of the basic protein that makeup the nucleosomes and have a fundamental role in chromosome structure.
- **HU protein:** The protein required to stimulate reaction leading to DNA replication.
- **Histocompatibility genes:** The gene that code for the histocompatibility antigens.
- **Histocompatibility antigens:** Antigen that determined the acceptance or rejection of a tissues graft.
- **SOS repair system:** Repair system (*rec A.* uvr) induced by the presence of single stranded DNA that usually occurs from post replicative gaps caused by various types of DNA damage.
- **SOS response:** The rec A protein stimulated by single stranded DNA is involved in the inactivation of the Lex A repressor as response.
- **Split gene:** A discontinuous or insertion sequences as known as intron.
- **Stable RNA:** A RNA molecule such as tRNA and rRNA but not mRNA.
- **Stacking:** The packaging of the flatish nitrogenous bases at the centre of DNA of double helix.
- **Stacking forces:** Weak hydrophobic forces among two adjacent base pairs of double helical DNA
- **Super coiling:** Different coiling state of DNA double helix.
- **Subvital gene:** A gene that causes the death of some proportion of all individuals that expresses it.
- **U1-U7 RNA:** A family of stable snRNA molecule originally found in rat Novikoff hepatoma cells but in advanced discovered in a range of mammals and other higher eukaryotes.
- **Unique DNA sequence:** A length of DNA with no repetitive DNA nucleotide sequence.
- **zDNA:** The DNA in which sugar content and phosphate linkage follow a *zig-zag* pattern. Such DNA is left handed double helix.

ANSWERS

Chapter 11: Gene (DNA, RNA and Alleles)

1. A	41. C	81. A	121. A	161. A	201. A	241. A
2. A	42. D	82. A	122. A	162. A	202. D	242. C
3. A	43. D	83. A	123. C	163. C	203. A	243. A
4. B	44. D	84. C	124. B	164. A	204. D	244. A
5. A	45. B	85. D	125. D	165. C	205. C	
6. D	46. C	86. A	126. C	166. A	206. A	
7. A	47. D	87. C	127. A	167. D	207. D	
8. A	48. A	88. A	128. D	168. A	208. A	
9. A	49. B	89. D	129. A	169. A	209. A	
10. D	50. B	90. A	130. B	170. C	210. A	
11. C	51. D	91. D	131. A	171. A	211. A	
12. D	52. A	92. A	132. A	172. A	212. A	
13. C	53. C	93. B	133. A	173. B	213. A	
14. A	54. A	94. A	134. C	174. A	214. A	
15. A	55. A	95. A	135. A	175. D	215. A	
16. D	56. A	96. C	136. A	176. A	216. A	
17. A	57. D	97. A	137. B	177. A	217. A	
18. A	58. A	98. D	138. C	178. A	218. C	
19. A	59. B	99. D	139. A	179. A	219. A	
20. A	60. A	100. D	140. A	180. A	220. C	
21. A	61. A	101. B	141. D	181. A	221. A	
22. C	62. D	102. A	142. A	182. A	222. A	
23. A	63. D	103. A	143. A	183. D	223. D	
24. A	64. A	104. A	144. A	184. B	224. A	
25. D	65. A	105. B	145. A	185. A	225. A	
26. C	66. A	106. A	146. A	186. A	226. D	
27. A	67. A	107. A	147. A	187. A	227. A	
28. A	68. A	108. A	148. A	188. C	228. A	
29. A	69. A	109. D	149. A	189. C	229. A	
30. A	70. A	110. D	150. C	190. D	230. A	
31. D	71. B	111. A	151. C	191. A	231. A	
32. B	72. A	112. A	152. A	192. D	232. D	
33. A	73. A	113. A	153. A	193. B	233. A	
34. B	74. A	114. A	154. D	194. B	234. D	
35. A	75. A	115. C	155. A	195. A	235. C	
36. A	76. A	116. A	156. B	196. A	236. A	
37. A	77. A	117. A	157. D	197. C	237. A	
38. A	78. D	118. A	158. A	198. C	238. B	
39. D	79. A	119. A	159. C	199. A	239. A	
40. D	80. A	120. A	160. A	200. A	240. B	

Chapter 12

Transcription and RNA Processing

1. Which of the following step is known as first step of gene expression?

   A. Translation B. Transcription

   C. Conjugation D. Termination

2. A process by which a DNA segment is converted into RNA by RNA polymerases is known as:

   A. Translation B. Transcription

   C. Nucleic acid D. Gene expression

3. Who described the protein synthesis by Central Dogma of Biology?

   A. Crick B. Watson

   C. Both A & B D. Stahl and Chase

4. Transcription process of gene expression occurs in :

   A. Nucleus B. Cytosol

   C. Cytoplasm D. E. R.

5. Which type of non genetic RNA is an active part of protein synthesis?

   A. rRNA B. tRNA

   C. mRMA D. All of these

6. The elongate DNA transcribed into RNA molecule encoded by at least one gene is called a:

   A. Transcription unit B. Translation

   C. Both A & B D. Transformation

7. Pre-rRNA is synthesized in which of the following?

   A. Nucleolus B. Cytosol

   C. E. R. D. All of these

8. Who observed the activity of RNA polymerase in 1960?
   A. Wises B. Barrel
   C. Volkin and Astracham D. F Jacob
9. Wiess observed the activity of RNA polymerase in one of the following animals:
   A. Rat liver B. Cow
   C. Frog D. None
10. How many RNA polymerase are required for the synthesis of all kind of RNAs in bacterial cell?
   A. One B. Three
   C. Four D. Five
11. RNA polymerase of prokaryotes consists of :
   A. Core enzyme B. Sigma factor
   C. Both A & B D. IF1
12. Which of the following unit of RNA polymerase is helpful in recognition of start signals of DNA molecule?
   A. Sigma factor B. Beta factor
   C. Omega D. Alpha
13. Core enzyme consists of:
   A. α unit B. β unit
   C. β' unit D. All of these
14. The β and β' unit of core enzyme are encoded by which of the following?
   A. rpoB and rpoC B. only rpoC
   C. tRNA D. None
15. Which of the following polypeptide is encoded by rpo*A*?
   A. β subunit B. β'unit
   C. α unit D. All of these
16. The sequence of amino acids in a polypeptide is known as :
   A. Primary structure of polypeptide B. Secondary structure of polypeptide
   C. Central dogma D. None
17. Tertiary structure of protein refers to :
   A. A segment of a polypeptide may exhibit a α helix
   B. The folding of polypeptide in three dimensional space
   C. The spatial interrelationship of the amino acids segments of the polypeptide
   D. All of these.
18. mRNA is synthesized in one of the following:
   A. Nucleus B. Cytosol
   C. Cytoplasm D. All of these

19. Which of the following unit of holoenzyme is involved in the initiation of transcription?
A. σ
B. α
C. β and β'
D. All of these

20. The complete RNA polymerase molecule is known as :
A. Core enzyme
B. Holoenzyme
C. Sigma unit
D. All of these

21. Holoenzyme contains which of the following?
A. Two alpha (α) unit or polypeptide
B. One β polypeptide and one β' polypeptide
C. One omega and one sigma polypeptide
D. All of these

22. Which of the following, step is outcome of transcription when sigma unit is released?
A. Initiation
B. Elongation
C. Termination
D. None

23. Which of the following enzyme is responsible for elongation of transcription in nucleus?
A. Core enzyme
B. Complete RNA polymerase
C. Holoenzyme
D. All of these

24. The consensus-10 sequences TATAAT is known as:
A. Pribnow box
B. Recognition site
C. Both A & B
D. None

25. TTGAGA refers to:
A. – 35 sequences
B. Recognition site
C. Both A & B
D. None

26. RNA polymerase I is located in which of the following?
A. Nucleolus
B. Nucleoplasm
C. Both A & B
D. Mitochondria

27. Which of the following enzyme is responsible to catalyze the synthesis of rRNA during transcription?
A. RNA polymerase I
B. RNA polymerase II
C. RNA polymerase III
D. DNA polymerase I

28. RNA polymerase I and II are found in which of the following?
A. Nucleolus
B. E. R.
C. Nucleoplasm
D. Chromatin

29. Which of the following enzyme is playing a key role in transcription?
A. DNA polymerase I
B. DNA polymerase II
C. Both A & B
D. RNA polymerase I

30. RNA polymerases do not respond which one activity:
    A. Replication of DNA
    B. Production of rRNA
    C. Production of tRNA
    D. All of these
31. Which is not true in respect of RNA polymerases?
    A. RNA polymerases are capable to initiating a new strand
    B. RNA polymerase synthesizes DNA
    C. Subunits contained in RNA polymerases must unwind the templates of DNA
    D. RNA polymerases are involved in transcription
32. Which of the followings is encoded by rpo*D?*
    A. Sigma factor
    B. Beta factor
    C. Omega factor
    D. All of these
33. In mammals, intermediate product of gene transcription is :
    A. rRNA
    B. tRNA
    C. mRNA
    D. hnRNA
34. Which of the following is responsible to synthesize small interfering RNA?
    A. RNA polymerase I
    B. RNA polymerase II
    C. DNA polymerase II
    D. RNA polymerase IV
35. Which of the following subunit of RNA polymerase has a site of ribo-nuclease activity?
    A. RPB II
    B. RPB III
    C. RPB IV
    D. None
36. RNA polymerase II is composed by which of the followings?
    A. Core and shelf mobile elements
    B. Clamp mobile element
    C. Jaw lope mobile elements
    D. All of these
37. In transcription, á amanitin acts on:
    A. RPB I subunit of RNA polymerase II
    B. RPB II subunit of RNA polymerase II
    C. RPB 12 subunit of RNA polymerase II
    D. None
38. Protein which is needed for transcription is known as:
    A. Transcription factors
    B. DNA polymerase I
    C. DNA polymerase II
    D. None
39. Homeobox genes code for homeodomain proteins in which of the following?
    A. Vertebrates
    B. Invertebrates
    C. Both A and B
    D. None
40. Homeoboxes were first discovered in :
    A. Drosophila
    B. Mice
    C. Frog
    D. Maize

41. High concentration of á-aminitin is expressed by its effect on which of the following?
    A. rRNA B. mRNA
    C. tRNA D. All of these
42. Which of the following RNAs forms large subunit of ribosome?
    A. 28*S* RNAs B. 5*S* RNAs
    C. 5.8*S* RNAs D. All of these
43. Holoenzyme is also known as:
    A. Holopolymerase B. Helicase
    C. DNA polymerase D. Ligase
44. Which of the following is responsible for transcription but is not a part of RNA polymerase?
    A. Transcription factors B. Ligase
    C. Both A & B D. None
45. Among the following enzyme which one is not involved in transcription:
    A. Primase B. Helicase
    C. DNA polymerase D. All of these
46. Which of the following sentences is true in respect of transcription in prokaryotes?
    A. Synthesis of RNA from DNA template
    B. The binding site of RNA polymerase lie within the promoter sequence
    C. The holoenzyme 1st binds at the promoter site and forms the closed promoter complex in initiation
    D. All of these
47. A region of DNA duplex being transcribed is known as:
    A. Transcription bubble B. Dna B
    C. RNAP I D. None
48. Which of the following enzyme is closely related with transcription?
    A. Topoisomerase I B. DNA gyrase
    C. Both A & B D. None
49. Which of the following activity begins when core enzyme acts in transcription?
    A. Elongation process B. Termination process
    C. Holoenzyme activation D. None
50. Which statement is not true for DNA transcription?
    A. Template strand and mRNA have complementary sequences
    B. Template strand is used as coding
    C. Transcription is in 5 prime to 3 prime direction
    D. Template strand and mRNA have indistinguishable sequences

51. Which of the following enzyme is accountable to major RNA synthesis?
A. RNA pol I
B. RNA pol II
C. RNA pol III
D. All of these

52. In most of the eukaryotes, transcription unit is:
A. Monogenic
B. Multigenic
C. Polygenic
D. Both B & C

53. Which of the following factor is involved in repair of damaged DNA?
A. TFIIH
B. RAP74 subunit of TFIIF
C. Both A & B
D. None

54. Which of the following is analogous to enhancers?
A. Upstream regulatory sequences
B. Downstream regulatory sequences
C. Both A & B
D. None

55. Immunoglobulin gene is an example of:
A. Enhancer within transcription unit
B. Fond only B lymphocites
C. Enhance the transcription
D. All of these

56. The downstream or internal promoter occurs in which of the following?
A. 5S RNA
B. tRNA
C. Both A & B
D. None

57. Who got Noble Prize in chemistry for studies of the molecular basis of eukaryotic transcription in 2006?
A. Jacob and Monod
B. Roger D. Kornberg
C. Walter Fiers
D. R. Ramakrishanan

58. Few proofreading mechanism come in which of the following event of transcription?
A. Elongation
B. Initiation
C. Termination
D. Pre-initiation

59. Template strand is known as:
A. Non coding strand
B. Coding strand
C. DNA helix
D. None

60. Ubiquitin is :
A. Universal polypeptide
B. Found in bacteria and mammal
C. A regulatory protein
D. All of these

61. Who discovered ubiquitin protein in eukaryotes in 1975?
A. Goldstein
B. Aaron Ciechanover
C. Albert Lasker
D. Avram Hershko

62. A process by which the introns are revoked:
A. RNA splicing
B. Spliceosome
C. Degradosome
D. mRNA

63. RNA is modified in :

A. Nucleus
B. Golgi bodies
C. E. R.
D. Ribosome

64. Which of the following RNA is transported to cytosol through nuclear pore?

A. Pre mRNA
B. tRNA
C. Mature mRNA
D. All of these

65. Which event is required to break exon-intron junctions and joining ends of the exons?

A. Splicing
B. Reverse transcription
C. Homeostasis
D. Transesterification

66. A short sequence just before the end of an intron at which the lariat intermediate is formed in splicing by joining the 5 nucleotide of the intron to the 2 position of an Adenosine:

A. Bench point
B. Ligation
C. Lariat
D. Spliced

67. How many types of snRNPs (small nuclear ribonucleoproteins) are involved in splicing of mRNA?

A. 5
B. 7
C. 8
D. 10

68. Function of cap in RNA processing is :

A. Increase the efficiency of translation
B. To provide the stability and protection
C. Assist slicing
D. All of these

69. The snRNPs assemble onto the pre-mRNA to form a large protein-RNA complex called:

A. Spliceosome
B. RNPs
C. Degradosomes
D. None

70. Regulation by means of a repressor protein that blocks transcription is referred to as :

A. Negative regulation.
B. Positive regulation
C. Successive regulation
D. None

71. Which of the following event occurs in RNA processing?

A. Trimming one or both of the ends of the primary transcript to the mature RNA length
B. Removing internal RNA sequences by a process called RNA splicing
C. Modifying RNA nucleotides either at the ends of an RNA or within the body of the RNA
D. All of these

72. Which of the following event occurs in nucleolus?

A. Splicing
B. Mature rRNA binds to ribosomal protein
C. Assembling of ribosome
D. All of these

73. Which of the following RNA is involved in RNA splicing?

A. snRNA
B. snoRNAs
C. Both A & B
D. rRNA

74. Which RNA is involved in rRNA cleavage and modification?

A. snoRNA
B. tRNA
C. snRNA
D. None

75. Transfer RNAs are transcribed individually from:

A. tRNA genes
B. rRNA genes
C. mRNA genes
D. snoRNA

76. Introns are spliced out of primary RNA a transcript by a large structure is known as:

A. Spliceosome
B. mRNA
C. tRNA gene
D. Degredosome

77. IgM gene is a:

A. Member of the immunoglobulin superfamily
B. An example of alternative splicing of a cellular gene in eukaryotes
C. Both A & B
D. None

78. Who discovered RNA splicing?

A. Phillip Allen Sharp and Richard J. Roberts
B. Kathrin and Joseph Walter Sharp
C. James Dewey Watson
D. Ann Holcombe

79. Which protein is essential for splicing?

A. SR protein
B. Dna B protein
C. Rec A protein
D. None

80. SR protein is not found in :

A. Unicellular organism
B. Multi-cellular organism
C. Both A & B
D. None

81. SR protein functions are:

A. Provide the stability of genome
B. Inhibit the formation of R loops in the DNA strand
C. Both A & B
D. None

82. mRNA processing follow which of the following steps?

A. Capping and splicing
B. Poly (A) tailing
C. RNA editing
D. All of these

83. Cleavage and poly Adenynation are tightly coupled in which of the following ?
   A. mRNA processing B. ssDNA processing
   C. dsDNA processing D. All of these
84. Removal of introns and ligation of exons is known as:
   A. Splicing B. RNA processing
   C. Transcription D. Translation
85. Splisosome consists of:
   A. 150 protein B. 5 rRNA
   C. Both A & B D. None
86. Kind of splicing is one which of the following ?
   A. Self splicing B. Trans splicing
   C. Alternative splicing D. All of these
87. Which of the following splicing responds to production of more than mRNA or protein traits?
   A. Self splicing B. Trans splicing
   C. Alternative splicing D. None
88. Two rRNA produce single mRNA molecule through :
   A. Self splicing B. Trans splicing
   C. Alternative splicing D. All of these
89. Which process transcribed DNA template from RNA ?
   A. Central dogma B. Reverse transcription
   C. Translation D. Replication

Terms

- **Transcription:** It is the first step of gene expression in which DNA is converted into different forms of RNA. In other hands, it is a process as transformation of genetic information from DNA to RNA.
- **Central Dogma of Biology:** A biological process carrying various steps like transcription, translation etc.
- **Pribnow box:** The most common or consensus -10 sequences as known as *TATAAT*.
- **Recognition sequences:** The consensus -35 sequences as known as *TTGAGA*.
- **CAAT box:** The consensus -75 sequences as known as *GGCCAATCT*.
- **Hogness box or TATA box:** The consensus -25 sequences as known as *TATAAAA*.

- **Spliceosome:** A product of splicing.
- **Insulators:** It is regulatory elements which shelter gene form appropriate regulatory interaction. It is also known as enhancer blocker and barrier as based their functional properties.
- **Initiator:** It is a conserved sequence encompassing the start site of bio-process.
- **Sense strand of DNA:** A DNA double helix has two strands of which one is called sense strand. This sense strand is used for transcription. It is also known as genic or coding strand of DNA.
- **Transcription factories:** A region where transcription by Pol III is also non-randomly and non-homogeneously distributed within the nucleus.
- **RNA replicases or RNA dependent RNA pol:** An enzyme use of RNA template for RNA synthesis.
- **Sigma factor:** A part of polymerase which associated with core enzyme responsible to synthesis of all kind of RNA (mRNA, tRNA, and rRNA) in *E. coli.*
- **RNA pol IV:** It is not a part of classical RNA polymerases and not involved in the synthesis of any RNA species during protein synthesis but also involve in transcription to lead the production of small RNA molecules that are associated with gene silencing.
- **Steroid receptors:** The receptors include receptors for the steroid hormones, retinoids, vit. D, thyroid hormones and other.
- **Activator domains of transcription:** A domain of transcription each consist of 300 amino acids.
- **Homeodomain:** It is a highly conserved domain of 60 amino acids found in a large family of transcription factor. This family was first identified in *Drosophila (i.e.*, legs of antina).
- **Inchworm model for elongation of transcripts:** It involves inchworm like translocation of RNA polymerase along the DNA template. RNAP is assumed to have one site for DNA binding and two sites for RNA binding (the 'leading product site' and the lagging product site')
- **Transcription terminators:** A DNA segment terminates the mRNA chain in prokaryotes. In other words, a section of nucleic acid sequence that marks the end of a gene or operon in genomic DNA during transcription.
- **Rho-dependent terminators:** Rho-dependent transcription terminators require a protein called Rho factor, which exhibits RNA helicase activity, to disrupt the mRNA-DNA-RNA polymerase transcriptional complex. Rho-dependent terminators are found in bacteria and phage.
- **Rho-independent terminators:** Intrinsic termination (also called Rho-independent termination) is a mechanism in prokaryotes that causes RNA transcription to be stopped.

- **Torpedo model:** The mRNA is completed, the residual RNA strand remains in association with the DNA template and the RNA polymerase II, continuing to be transcribed. XRN2 (5'-3' Exoribonuclease 2), a RNase, attaches to the carboxyl terminal domain of RNA polymerase II and proceeds to degrade the uncapped residual RNA from 5′ to 3′ until it reaches the RNA pol II. 5' cap refers to a modified guanine added to the front of mRNA for protection from RNase.
- **Allosteric model:** RNA polymerase normally is capable of transcribing DNA into single-stranded mRNA efficiently. However, upon transcribing over the poly-A signals on the DNA template, a conformational shift is induced in the RNA and odiase polymerase from the proposed loss of associated proteins from its carboxyl terminal domain.
- **Reverse transcriptase:** A Reverse transcriptase (RT) is an enzyme used to generate complementary DNA (cDNA) from an RNA template.
- **cDNA library :** A cDNA library is a combination of cloned cDNA (complementary DNA) fragments inserted into a collection of host cells, which together constitute some portion of the transcriptome of the organism.
- **Multicopy single-stranded DNA:** msDNA is a type of extrachromosomal satellite DNA that consists of a single-stranded DNA molecule covalently linked via a 2'-5' phosphodiester bond to an internal guanosine of an RNA molecule.
- **Snurp (small nuclear RNA protein):** It is a type of protein use in post transcription.
- **Exosome:** It is a multi-protein complex capable of degrading various types of RNA (ribonucleic acid) molecules. Exosome complexes are found in both eukaryotic cells and archaea, while in bacteria a simpler complex called the degradosome carries out similar functions.
- **Core proteins:** The core of the complex has a ring structure consisting of six proteins that all belong to the same class of RNases, the RNase PH-like proteins.
- **Splicing:** It is a modification of the nascent pre-messenger RNA (pre-mRNA) transcript in which introns are removed and exons are joined. For nuclear encoded genes, splicing takes place within the nucleus after or concurrently with transcription.

ANSWERS

Chapter 12: Transcription and RNA processing

1. B	21. D	41. D	61. A	81. C
2. B	22. B	42. D	62. A	82. D
3. A	23. A	43. A	63. A	83. A
4. A	24. A	44. C	64. C	84. A
5. D	25. C	45. D	65. A	85. C
6. A	26. A	46. D	66. A	86. D
7. A	27. A	47. A	67. A	87. C
8. A	28. C	48. C	68. D	88. B
9. A	29. D	49. A	69. A	89. B
10. A	30. A	50. D	70. A	
11. C	31. B	51. D	71. D	
12. A	32. A	52. A	72. D	
13. D	33. A	53. C	73. A	
14. A	34. D	54. A	74. A	
15. C	35. A	55. D	75. A	
16. A	36. D	56. C	76. A	
17. B	37. A	57. B	77. C	
18. A	38. A	58. A	78. A	
19. A	39. C	59. A	79. A	
20. B	40. A	60. D	80. A	

Chapter 13

Genetic Code and Translation

1. The group of all codon used to identify 20 amino acids considered as which of the following:

   A. Genetic code B. Genetic language

   C. Coding directory D. All of these

2. How many letter carried by each codon?

   A. Three B. Four

   C. Two D. One

3. Codon is not code of any amino acids is known as:

   A. Nonsense codon B. Sense codon

   C. Redundant D. Regenerate

4. The mRNA nucleotides are the nucleotides sequences that code to a particular amino acid is called:

   A. Codon B. Ambiguous

   C. Regeneration D. Redundant

5. Who coined the term codon in 1963?

   A. Brenner *et al.*, B. Crick

   C. Beadle and Tatum D. Gamow

6. Who postulated the sets of three bases must be employed to encode the 20 standard amino acids used by living cells to build proteins in 1954?

   A. George Gamow B. Brenner *et al.*,

   C. Crick and Watson D. Beadle and Tatum

7. Properties of genetic code is one of the following:

   A. Code is triplet and degenerate B. Code is non overlaps and comma less

   C. Code is non ambiguous and universal D. All of these

8. Who demonstrated that codon consist of three DNA bases in 1961?
   A. Crick
   B. Brenner *et al.*,
   C. Both A & B
   D. Beadle
9. Who synthesized RNA using only uracil?
   A. M. W. Nirenberg
   B. J. H. Mattaei
   C. Both A & B
   D. R. W. Holley
10. Chain terminating codon or nonsense codons are:
   A. UAA
   B. UGA
   C. UAG
   D. All of these
11. Genetic codon was discovered in:
   A. 1960
   B. 1963
   C. 1978
   D. 1999
12. The relationship between the sequence of nucleotides of DNA and the sequence of amino acid in proteins is known as:
   A. Genetic code
   B. Translation
   C. Transcription
   D. Transformation
13. Genetic code written in which of the following direction
   A. 5 prime to 3 prime
   B. 3 prime to 5 prime
   C. Both A & B
   D. None
14. The difference in the frequency of occurrence of synonymous codons in coding DNA is known as:
   A. Codon usage bias
   B. Genetic codon
   C. Genetic language
   D. None
15. Which of the factor to be contributing codon usage bias?
   A. Gene expression level
   B. GC composition
   C. Protein hydropathy
   D. All of these
16. Who proposed Wobble hypothesis in 1965?
   A. F. H. C. Crick
   B. T. H. Morgan
   C. Barrel
   D. Abelson
17. Guthrie and Abelson proposed the revise wobble hypothesis in 1982 is based on:
   A. Characterization of tRNA
   B. tRNA anti codon
   C. Both A & B
   D. None
18. Which hypothesis represented codon reassignment?
   A. Codon capture hypothesis
   B. Ambiguous intermediate hypothesis
   C. Genome streamlining hypothesis
   D. All of these

19. Who presented the concept of primitive code in 1966?

A. T. H. Jukes B. Crick *et al.*,

C. Abelson D. Nirenberg

20. NCBI stands for:

A. National Centre for Biotechnology Information

B. National Citrus Biotech Institute

C. National Bureau of Citrus information

D. None

21. Which factor is responsible to premature termination of polypeptide chain?

A. Nonsense mutation B. Nonsense codon

C. Suppressor gene D. All of these

22. The codon AUG acts as:

A. Initiator codon B. Nonsense codon

C. Terminator codon D. All of these

23. Wobble hypothesis explained

A. Only degeneracy of codon at 3rd position of codon

B. Ambiguity in the genetic codon

C. Both A & B

D. Commaless of codon

24. Sequences of tRNA which pairs with the sequences of mRNA during translation is known as:

A. Anticodon B. Protein synthesis

C. Genetic codon D. None

25. Which direction is used for reading the massages in mRNA:

A. 5 prime to 3 prime B. 3 prime to 5 prime

C. Both A & B D. None

26. Who proposed the Wobble base pairing in 1966?

A. Crick B. Watson

C. Both A & B D. None

27. Which nucleotide is newly discovered?

A. Inosine B. Ochre

C. Amber D. None

28. Which of the following amino acid is initiate the translation in prokaryotes?

A. Methionine B. N-formyl Methionine

C. Inosine D. Leusine

29. Which amino acid initiates the translation in eukaryotes?
    A. Methionine B. N-formyl Methionine
    C. Inosine D. Leusine
30. Who advised triplet codon in 1954?
    A. Brenner *et al.*, B. Crick and Watson
    C. Beadle and Tatum D. Gamow
31. A process by which the conversion of mRNA base sequences into protein amino acid sequences is known as:
    A. Transcription B. Transduction
    C. Special transduction D. None
32. Translation occurs in:
    A. Cytoplasm B. Nucleus
    C. Both A & B D. None
33. Translation is a process by which:
    A. mRNA base sequences converted into protein amino acid sequences
    B. Protein amino acid sequences converted into mRNA base sequences
    C. Protein degradation
    D. None
34. Translation of mRNA is required for which of the following species of RNA:
    A. rRNA B. tRNA
    C. Both A & B D. None
35. Regulatory gene is found in:
    A. Prokaryotes B. Eukaryotes
    C. Both A & B D. None
36. Which process create protein from ribosome?
    A. Translation B. Transcription
    C. Transduction D. Special transcription
37. Which of the following phases come in translation?
    A. Initiation B. Elongation
    C. Termination D. All of these
38. In eukaryotes, translation occurs in cytosol or across the membrane of the endoplasmic reticulum is known as:
    A. Vectorial synthesis B. Secretion
    C. Transmission D. None
39. Which of the following antibiotics act by inhibiting translation?
    A. *Acycloheximide* and *tetracycline* B. *Chloramphenicol* and *streptomycin,*
    C. *Erythromycin* and *puromycin* D. All of these

40. The ribosome is a multi subunit structure containing which of the following:
    A. rRNA B. Proteins
    C. Both A & B D. None

41. In prokaryotes, which of the following contains information for the amino acid sequences for different polypeptide chains is known as:
    A. mRNA B. DNA
    C. cDNA D. cpDNA

42. Which feature is closely linked with translation?
    A. RNA is translated from the 5 prime end of the molecule towards the 3 prime end
    B. Polypeptides are synthesized from the amino terminus towards the carboxyl terminus, by adding amino acids one by one to the carboxyl end.
    C. Both A & B
    D. None

43. Pre- initiation of translation includes one of the following:
    A. Formation of aminoacyl tRNA B. Dissociation of ribosome into its two subunits
    C. Both A & B D. None

44. Which of the following step is involved in the formation of aminoacyl tRNA?
    A. Activation of amino acid B. Transfer of amino acid to tRNA
    C. Both A & B D. None

45. Which of the following factor is responsible to dissociation of ribosome into its two subunits in prokaryotes?
    A. IF1 B. IF2
    C. IF3 D. All of these

46. Which of the following factor is responsible to dissociation of ribosome into its two subunits in eukaryotes?
    A. eIF6 B. eIF 1
    C. eIF 4e D. None

47. Which of the following codon is initiated the translation rarely in prokaryotes?
    A. GUG B. UAA
    C. UAG D. None

48. Which of the following factor is facilitated the formation of initiation complex?
    A. IF1 B. IF2
    C. IF3 D. All of these

49. Shine Delgarno sequences of mRNA is located at:
    A. Upstream of translation sequence B. Downstream translation sequence
    C. Both A & B D. None

50. Shine Delgarno sequences are complementary of one of the following:

A. 3 prime of 16S rRNA
B. tRNA
C. mRNA
D. AUG

51. Anti Shine Delgarno region is found in which of the following;

A. 30S subunit of ribosome
B. 40S subunit of ribosome
C. Both A & B
D. None

52. Which of the following is essential for binding of the anti-codon stem loop of tRNA to P site of the small subunit of ribosome?

A. 16S rRNA
B. 23S rRNA
C. 80S RNA
D. All of these

53. Initiation complex of in translation made by which of the following:

A. 70S RNA
B. mRNA
C. f-met-$tRNA_f^{met}$
D. All of these

54. Elongation of translation, which energy source is used:

A. GTP
B. ATP
C. Both A & B
D. None

55. Which factor play a key role for the detection of the correct initiation codon in eukaryotes?

A. eIF1
B. eIF6
C. eIF3A
D. None

56. Which factor is responsible to hydrolysis of GTP in translation?

A. eIF2 in eukaryotes
B. IF2 in prokaryotes
C. Both A & B
D. None

57. The first amino acid incorporated into proteins:

A. can be any of the 20 standard amino acids
B. is an N-formylmethionine in *E. coli* and methionine in other organisms
C. is always inosinate
D. is an amidated methionine residue that is cleaved following termination of translation

58. How many GTP are hydrolyzed for every aminoacyl-tRNA that is successfully inserted into the A site of the ribosome?

A. zero
B. 1
C. 2
D. 4

59. At the end of the initiation step, the ribosome has a vacant:

A. A site
B. P site
C. mRNA site
D. tRNA

60. How are the termination codons different from other codons?

A. They contain thymines.

B. The termination codon always codes for methionine.

C. They are not recognized by any tRNA molecules.

D. Their conformations do not allow them to fit properly in the A site of the ribosome.

61. In translation, GTP hydrolysis is used for all of these except:

A. Position the initial aminoacyl-tRNA in the ribosome

B. Catalyze peptide bond formation

C. Position incoming aminoacyl-tRNAs into the ribosome for elongation

D. Provide energy for translocation of the ribosome to the next codon.

62. The 50S sub unit of ribosome is composed of:

A. 23S rRNA　　B. 5S rRNA

C. 31 different protein　　D. All of these

63. A complex of 16S rRNA and 21 different proteins is known as:

A. 30S subunit of ribosome　　B. 50S subunit of ribosome

C. 60S subunit of ribosome　　D. All of these

64. Which of the following factor is required to hydrolysis the Gamma Phosphate of the bound GTP, converting it to GDP?

A. IF2　　B. IF1

C. IF3　　D. All of these

65. Which of the following factor is elongate the translation?

A. EF-G　　B. IF 1

C. eIF4　　D. None

66. A small subunit (30S) of ribosome interacts with proteins is called as:

A. Initiation factor　　B. Elongation factor

C. Termination factor　　D. All of these

67. Which of the following factor is known as release factor used for termination in translation?

A. RF1　　B. RF2

C. Both A & B　　D. GUG

68. GTP bound by RF-3 is hydrolyzed to

A. GDP　　B. ATP

C. ADP　　D. None

69. Who suggested the Kozak's ribosome scanning hypothesis for translation in eukaryotes in 1983?

A. Jacob　　B. T.F.C. Makay

C. Both A & B　　D. None

70. Which of the following elongation factor is required for translocation in eukaryotes?

A. eEF 2
B. eEF4
C. Both A & B
D. None

71. Which of the following codon is recognized by RF1 in prokaryotes?

A. UAA
B. UAG
C. Both A & B
D. UGA

72. Which mechanism deals with information stored in the sequences of nucleotides in the mRNA molecules converted into the sequences of amino acids in the protein gene products?

A. Translation
B. Transcription
C. Reverse transcription
D. None

73. Who proposed the alkaptonuria disease caused inherited by single recessive gene?

A. Garrod
B. Mendel
C. Morgan
D. Muller

74. Antibiotic streptomycin inhibit the translation in prokaryotes by:

A. Prevent the initiation
B. Bind with 30 S subunit and stabilize the entire initiation complex
C. Stop the jointing 50 S subunit
D. Inhibit the elongation

75. Ribosome is the workshop of:

A. Protein biosynthesis
B. Primase
C. DNA biosynthesis
D. DNA polymerase

76. Which of the following carries the blue print of life?

A. mRNA
B. rRNA
C. miRNA
D. snRNA

77. Which event is not required any enzymatic reaction?

A. Binding of tRNA to ribosome
B. Synthesis of mRNA
C. Translation
D. None

78. Along the helix each amino acid residue advances by:

A. 0.14 nm
B. 0.20 nm
C. 0.15 nm
D. None

79. Which is not the trait of a codon?

A. Degeneracy
B. Universal
C. Ambiguous and overlapping
D. Continuous and commaless

80. In the biosynthesis of c-DNA, the joining enzyme ligase requires

A. GTP
B. ATP
C. CTP
D. UTP

81. The 'rho' (ρ) factor is involved in which of the following:
   A. To increase the rate of RNA synthesis
   B. In binding catabolite repressor to the promoter region
   C. In proper termination of transcription
   D. To allow proper initiation of transcriptide

82. Which of the following differ from each other?
   A. Opal B. Valine
   C. Methionine D. None

83. Valine difference to opal is based on:
   A. It is a nonsense codon rarely
   B. Its terminate the polypeptide chain
   C. Both A & B
   D. None

84. Which one is different :
   A. GUG B. UUU
   C. AUG D. UAG

85. Which of the following is different :
   A. Opal B. Amber
   C. Ochre D. Leucine

86. Who synthesized the gene by chemical method in 1970?
   A. Khorana B. Morgan
   C. Correns D. None

Terms

- **Anti-codon:** Sequences of tRNA which pairs with the sequences of mRNA during translation is called anti-codon.
- **Antisense RNA:** A RNA transcribed by using antisense DNA strand.
- **Code:** A sign encode specific information.
- **Cognate tRNA:** tRNA that are recognized by a particular aminocyl-tRNA synthetase.
- **Genetic code:** A group of RNA bases written in triplet form with direction as 5' to 3'; encoded specific information in respect to inheritance or character expression. In other words, group of the all codon use to identifying 20 amino acids as considered genetic code or genetic language or code directory. Example, AUG is a genetic code encoded the information about initiation of bio-pathway.

- **Neutrality theory of protein evolution:** Rates of amino acid replacement in proteins and nucleotide substitution in DNA during evolution may be approximately constant because the vast majority of such changes are selectively neutral.
- **Nick translation:** The ability of *E. coli.* DNA polymerase I to use a nick as a starting point from which one strand of a duplex DNA can be degraded and replaced by resynthesis of new material.
- **Non ambiguous codon:** A nature of codon where one codon code only one amino acid.
- **Non genetic material:** The RNA which is found in addition to DNA and does not act as genetic material.
- **Non overlapping codon:** Nature of genetic codon where one nucleotide is part only one code.
- **Nonsense codon:** A sequence of mRNA *i.e.*, codon signals the termination of chain of translation. These codon are UAA, UAG, and UGA.
- **Peptidyl transferase**: An enzymatic activity responsible for peptide bond synthesis during translation.
- **Polycistronic mRNA:** A type of mRNA found in mostly prokaryotes that contain information from more than one amino acid.
- **Pribnow box:** Closely invariable six nucleotide DNA sequence TATAAT, located in eukaryotes promoters upstream (-10bps) from start codon.
- **Primer:** A short sequences of nucleotide with a reactive 3'-OH that can initiate DNA synthesis along template.
- **Primosome:** A complex of two proteins (primase and helicase) that initiate RNA primers on lagging DNA strand during DNA replication.
- **Recognition site:** A sequence of bases where promoter is present and serves to several recognize RNA polymerase molecule.
- **RNA polymerase:** The enzymes that catalyzed RNA synthesis during transcription in *E. coli.*
- **Shine Dalgarno sequence:** The prokaryotic ribosome binding site AGGAGG presented on mRNA just prior to AUG codon which has complementary with the 3'end of 16 rRNA.
- **Splicing:** The process when covalent joined exon sequences of RNA and discard the introns sequences.
- **Stable RNA:** Such rRNA and tRNA.
- **TACTAAC box:** A consensus sequence covered by lariat branch point of eukaryotic mRNA.
- **TATA box:** A group of nucleotide presented on -25 bps that makes the prokaryotic promoter.
- **Translation:** A process by which mRNA base sequences converts into protein. It occurs in cytoplasm.
- **Vectorial synthesis:** In eukaryotes, translation occurs in cytosol or across the membrane of the endoplasmic reticulum.
- **Wobble base pairing:** The pairing of mRNA codon with tRNA anticodon in which first two base of a codon have normal base pairing and third base have abnormal base pairing.

ANSWERS

Chapter 13: Genetic code and translation

1. D	21. D	41. A	61. B	81. C
2. A	22. A	42. C	62. D	82. A
3. A	23. A	43. C	63. A	83. A
4. A	24. A	44. C	64. A	84. D
5. B	25. A	45. A	65. A	85. D
6. A	26. A	46. A	66. A	86. A
7. D	27. A	47. A	67. C	
8. C	28. B	48. D	68. A	
9. C	29. A	49. A	69. D	
10. D	30. D	50. A	70. A	
11. A	31. D	51. A	71. C	
12. A	32. A	52. A	72. A	
13. A	33. A	53. D	73. A	
14. A	34. C	54. A	74. B	
15. D	35. C	55. A	75. A	
16. A	36. A	56. C	76. A	
17. C	37. D	57. B	77. A	
18. D	38. A	58. C	78. C	
19. A	39. D	59. A	79. C	
20. A	40. C	60. C	80. B	

Chapter **14**

Molecular Marker and Biotechnology

1. Which of the following is mapped as a number of restriction site of a restriction enzyme in DNA molecule?

   A. Restriction mapping B. Chromosome walking

   C. Both A & B D. None

2. Restriction sites are separated by their actual distance measured by which of the following unit?

   A. Nucleotide base pairs B. cM unit

   C. Recombination frequencies D. All of these

3. The unit of linkage mapping or genetic map is :

   A. Recombination frequency B. Nucleotide base pairs

   C. Both A & B D. None

4. Both physical and genetic maps are available in which of the following?

   A. Maize, potato and tomato B. Rice, wheat and barley

   C. *Drosophila,* mouse and human D. All of these

5. Which is not true in respect of recombination frequency?

   A. Use for measuring the distance between two genes

   B. Influenced by nature of mutant, position of mutant on chromosome

   C. Not affected by environment condition and genetic background of organism

   D. None

6. Fine structure of a gene can be studied with the help of

   A. Determination of the sequence of nucleotide of concerned DNA molecule

   B. Recombination frequency

   C. cM measurement unit

   D. All of these

7. Restriction map does not linked with features of the following:
   A. A strength line representing a DNA segment of defined length
   B. Mapped by recombination frequency
   C. Restriction site represented by dots
   D. The distance between restriction site mapped in base pairs
8. Restriction mapping is a technique successfully used in which for mapping restriction sites in DNA segment:
   A. Eukaryotes B. Prokaryotes
   C. Both A & B D. None
9. The large changes (cause) on the genetic map can be successfully located on restriction mapped except:
   A. Deletion B. Duplication
   C. Inversion and intercganges D. Point mutation
10. Which of the following is produce of mapping population?
   A. F_1 B. BC_1
   C. F_2 D. All of these
11. Which can be maintained easily due to homozygosity in population?
   A. RILs B. DHs
   C. Both A & B D. F_1
12. Which of the following marker is known as co-dominant marker?
   A. RFLP and RAPD B. SSR/STR
   C. SNP D. All of these
13. Which one is different than other :
   A. RFLP B. SSR
   C. SNP D. RAPD
14. Which of the following marker is dominant marker?
   A. AFLP B. SSR
   C. SNP D. All of these
15. Which marker does not require PCR technique?
   A. RFLP B. SNP
   C. SSR D. SCAR
16. Which hybridization marker is based used is for preparation of genetic map of gene?
   A. RFLP B. SSR
   C. STR D. SNP
17. Which marker is based on PCR as well as on microarray?
   A. SNP B. SSR
   C. AFLP D. STS

18. Which of the following is first generation marker?
A. RFLPs B. RAPDs
C. Both A & B D. SNP

19. Which one of the following is third generation marker?
A. SNP B. SSR
C. STS D. AFLPs

20. Second generation marker is:
A. SSR, STS, AFLPs and STS B. SNP
C. RFLPs D. RAPDs

21. Forth generation marker is :
A. SFPs B. DArT
C. Both A & B D. None

22. Array based molecular markers are considered as :
A. Forth generation marker B. Second generation marker
C. Third generation marker D. None

23. Which characteristic is not based on RFLPs ?
A. Hybridization B. Co-dominant
C. Length polymorphism D. PCR

24. Which of the following polymorphic marker used first for genetic mapping in human being?
A. RFLPs B. SNP
C. SSR D.. STS

25. Which of the following feature does not based on RAPD marker?
A. PCR
B. Amplifies inverted repeats in the genome
C. Can generate larger number of markers than RFLP in a short period of time
D. Require radioisotopes

26. Micro satellite is also known as:
A. SSR = Simple Sequence Repeats
B. SSLP = Simple Sequence Length Polymorphisms
C. STMS = Sequence Tagged Microsatellites
D. All of these

27. RAPD joints which of the following molecular marker, resulting increase the efficiency for genetic analysis?
A. RFLPs B. SRR
C. STS D. None

28. RAPD has been found to be incompetent due to its which feature?
    A. Low reproducibility
    B. Same primers sequence with arbitrary may be used for different species
    C. No specific probe required
    D. All of thses
29. Which of the following is used for DNA fingerprinting in human being?
    A. Mini-satellites B. SSRs
    C. Both A & B D. None
30. The DNA profiling technique was first reported in 1986 by :
    A. Alec Jeffreys B. William M. Bass
    C. George W. Gill D. Edmond Locard
31. Which property is abscent in DAFs.
    A. Higher primer concentration are used in DAF
    B. Same temperature cycles in DAF and RAPD
    C. Banding pattern is very complex
    D. Product of DAF is detected by silver staining
32. Which is not true in respect of AFLPs?
    A. Highly reproducibility capacity B. PCR based
    C. Novel DNA fingerprinting approach D. None
33. Who coined the term microsatellites in 1989?
    A. Aec Jaffrey B. Litt and Lutty
    C. Willium, J. D. None
34. Which of the following is ideal marker compared to other for genetic mapping?
    A. RFLP B. RAPD
    C. SSR D. None
35. Which of the following approach is known as single primer amplification reaction?
    A. Microsatellite directed PCR (unanchored primers) technique
    B. Microsatellite directed PCR (anchored primers) technique
    C. Both A & B
    D. RAMP
36. Which technique is used for dominant and co-dominant polymorphism fragment of genome?
    A. Random amplified microsatellite polymorphism (RAMP)
    B. Selective amplification of microsatellite polymorphic loci (SAMPL)
    C. RFLP
    D. All of these

37. Retrotransposon is most important automobile for:
A. Chromosome rearrangement
B. Gene duplication
C. Mutagenesis
D. All of these

38. Who described Sequence-specific amplified polymorphism marker in 1997?
A. Waugh *et al.*,
B. Aec Jaffrey
C. Litt and Lutty
D. None

39. Which of the following marker created the bridges between retro-transposons and non retrotransposon fragment of genome?
A. S-SAP
B. RAMP
C. SAMPL
D. SNP

40. Which of the following marker is required only one restriction enzyme?
A. AFLP
B. SAMPL
C. Both A & B
D. None

41. The exploitation of retrotransposons as marker in plant was first described in 1990 by scientist:
A. Lee *et al.*,
B. Waugh *et al.*,
C. Melayah *et al.*,
D. None

42. Which of the following marker is not inherited in both alleles of a gene when only epistatic allele mapped?
A. SCARS (Sequence characterized amplified regions)
B. RFLPs
C. SSR
D. RAPDs

43. PCR-RFLP marker is known as:
A. Cleaved Amplified Polymorphic Sequences
B. Inter Retrotransposon Amplified Polymorphism
C. Both A & B
D. Selective amplification of microsatellite polymorphic loci (SAMPL)

44. Which approach is widely used for DNA fingerprinting of an individual?
A. RFLP
B. SSR
C. STS
D. SNP

45. Which is known as classical DNA fingerprinting derivatives of the following?
A. RFLP
B. RAPD
C. SNP
D. None

46. Which of the following types of probe applied to RFLP fingerprinting in numerous animal and plant species?
A. Minisatellites proble
B. Micro satellites probe
C. Both A & B
D. None

47. DNA fingerprinting technique is used for:
    A. Genome analysis
    B. Paternity testing
    C. Genotypes identification and population genetics
    D. All of these
48. Centre for DNA fingerprinting and Diagnostics is situated at:
    A. Hyderabad
    B. Nagpur
    C. Bangalore
    D. Delhi
49. Gene replaced by normal functional gene is based on which of the following theory?
    A. Gene theory
    B. Lyon theory
    C. DNA fingerprinting
    D. All of these
50. Which of the following is used for gene theory?
    A. Somatic patient gene
    B. Germline gene
    C. Both A & B
    D. None
51. The process by which genes are cloned called:
    A. Forward mutagenesis
    B. Reverse mutagenesis
    C. Reverse transcription
    D. Reverse genetics
52. Reverse genetics involves which of the following step?
    A. Locating the gene to a particular position on chromosome via RFLP linkage analysis
    B. Map the position of the gene with respect to molecular marker
    C. Both A & B
    D. None
53. rRNA can be:
    A. 28 S and 16S
    B. 5.8S
    C. 5S
    D. All of these
54. Who isolated rRNA genes from amphibian (*Xenopus*) ?
    A. H. Wallace and M.L. Birnstiel
    B. Morgan
    C. Donald D Brown
    D. None
55. rRNA genes belongs to:
    A. Multigenes family
    B. Single gene family
    C. Both A & B
    D. None
56. Complementary DNA (cDNA) may be used for:
    A. Direct gene manipulation
    B. Transformation
    C. Both A & B
    D. Conjugation
    E. None

57. The mRNA is used for copying into cDNA by which of the following process?

A. Transformation
B. Transfection
C. Translation
D. Reverse transcription

58. Which RNA is smallest RNA?

A. rRNA
B. tRNA
C. mRNA
D. None

59. Which RNA is longest RNA found in cell?

A. mRNA
B. rRNA
C. tRNA
D. None

60. Reporter gene is attached by which RNA for identification of transformants:

A. tRNA
B. snRNA
C. mRNA
D. None

61. The Probe obtained from one plant species and used for another plant species in cloning is known as:

A. Homologue probe
B. Heterologous probe
C. Partial probe
D. None

62. Hargobind Khurana delivered his Noble lecture on :

A. December, 12, 1968
B. November, 12, 1968
C. July, 12, 1968
D. None

63. Which of the following gene is derived from reverse transcription?

A. Dominant gene
B. Over-dominance gene
C. Pseudogene
D. None

64. Pseudo-gene derived from which of the following material?

A. mRNA
B. tRNA
C. rRNA
D. All of these

65. Who coined the term biotechnology in 1919?

A. Karl Ereky
B. Kotte
C. Haberlandt
D. White

66. A process of biological science in which direct or indirect use of part of living organism in artificially or natural is called:

A. Biotechnology
B. Genetic engineering
C. Gene manipulation
D. All of these

67. Which biological application involved the integrated use of biochemistry, microbiology and engineering science?

A. Biotechnology
B. Genetics
C. Molecular genetics
D. Genomics

68. Who was known as father of plant tissue culture?
    A. Gottelieb Haberlandt B. Kotte
    C. Hanning D. None

69. Who developed the concept of in vitro cell culture?
    A. Gottelieb Haberlandt B. Kotte
    C. Hanning D. None

70. Who recovered the virus free Dahilia by shoot meristem technique in 1952?
    A. Haberlandt B. Hanning
    C. Morel and Martin D. H. Hu

71. Who laid the foundation of commercial plant tissue culture in 1962?
    A. Morel, G.M. B. Haberlandt, G.
    C. Both A & B D. Martin

72. Auxins inhibit which of the following developmental activity in plant:
    A. Callus formation B. Cell division
    C. Cell elongation D. Adventitious and axillary shoot formation

73. Movement of auxins in plant is:
    A. Base to top B. Top to base
    C. Both A & B D. None

74. Which hormone exists antagonist of rhizogenesis?
    A. Auxins B. NAA
    C. Cytokinin D. GA3

75. Which hormones have rhizonic effect?
    A. IAA B. NAA
    C. IBA D. All of these

76. Which hormone is considered as cytokinin class?
    A. Naphthalene acetic acid B. Picloram
    C. Zeatin D. Indole acetic acid

77. Which hormones is synthesized naturally in plant?
    A. Zeatin B. 2-iP
    C. Both A & B D. None

78. Gibberellins inhibits which of the following developmental activity in plant:
    A. Adventitious root formation B. Shoot formation
    C. Both A & B D. Plant regeneration

79. Which unit used for measuring the lights?
    A. Foot candles B. Joule/meter square
    C. Both A & B D. None

80. Which range of pH of culture medium suitable for explants invitro culture?

A. 5.0 to 7.0 B. 4.0 to 6.5

C. Both A & B D. 4.0 to 8.0

81. Which process is used for elimination of microorganisms?

A. Sterilization B. Vertification

C. Turbidostat D. None

82. Which temperature is most preferable for autoclaving media?

A. 121 °C with pressure 15 psi for 20 minute

B. 111 °C with pressure 15 psi for 20 minute

C. 160 °C with pressure 15 psi for 20 minute

D. 100 °C with pressure 15 psi for 20 minute

83. Tissue culture refers to:

A. Invitro cultivation of explants B. Invivo cultivation

C. Both A & B D. None

84. Which phenomenon exists when a mature cell revert to a meristem and forming undifferentiated callus is known as?

A. Dedifferentication B. Absorption

C. Reversion D. None

85. A mass of undedifferentiated tissue is known as:

A. Petiolate B. Callus

C. Leaf sheath D. None

86. A piece of plant have potential to differentiate into shoot and root is called:

A. Callus B. Totipontency

C. Prepotency D. None

87. Clonal culture in vitro condition is also known as:

A. Micro propagation B. Cell culture

C. Clonal breeding D. None

88. Who coined the term clone?

A. Webber B. Went

C. Kotte D. Haberlandt

89. The term Clone is derived from which language:

A. Latin B. Germany

C. Greek D. British

90. Explants take which of the following?

A. Angiosperm B. Gymnosperm

C. Pteridophytes D. All of these

91. Which of the following is major objective of meristem culture?
    A. To develop disease free plant
    B. Multiplication of clone
    C. Creation of somatic variation
    D. All of these
92. Size of meristem may be frequently used for development of virus free plant:
    A. 0.2 to 0.5 mm
    B. 1.0 to 1.50 cm
    C. 0.80 to 0.9 mm
    D. None
93. Recommended ratio between cytokinin and auxins is as
    A. 10:1
    B. 10:5
    C. 10:3
    D. None
94. Who was first recognized the initiation and development of embryos from somatic cell in vitro in 1958?
    A. Steward *et al,*
    B. Canon
    C. White
    D. None
95. Which cell suspension culture has fixed volume of nutrients culture media?
    A. Batch culture
    B. Continuous culture
    C. Semi continuous culture
    D. None
96. Which method is used for achieving the synchronization of cell suspension culture?
    A. Mitotic index
    B. Starvation method
    C. Colchicines treatment
    D. All of these
97. Which is not DNA synthesis inhibitors?
    A. 5-fluorodeoxyuridine
    B. Thymidine
    C. Hydroxyurea
    D. None
98. Number of cells per volume of culture medium is called :
    A. Plating density
    B. Culture density
    C. Pore density
    D. None
99. Passaging cells is also known as:
    A. Subculture cells
    B. Splitting cells
    C. Both A & B
    D. None
100. Which method used for manipulating cells involves in the introduction of foreign DNA?
    A. Transduction
    B. Transfection
    C. Both A & B
    D. Plating density
101. Cell culture is a fundamental component of :
    A. Tissue engineering
    B. Tissue culture
    C. Both A & B
    D. Batch culture
102. Which of the following is Plant cell culture method?
    A. Insect cell culture
    B. Bacterial and yeast culture method
    C. Viral culture method
    D. All of these

103. The egg cell is always :
A. Asymmetric B. Symmetric
C. Both A & B D. None

104. Fertilization is also known as :
A. Conception B. Fecundation
C. Syngamy D. All of these

105. Which process exists the development of embryo by asymmetric cell division?
A. Embyrogenesis B. Androgenesis
C. Tissue culture D. None

106. Embryogenesis consists of the following phase:
A. Morphogenetic events B. Postembryonic development
C. Both A & B D. Cell division

107. Which is true in respect of zygotic embryo formation?
A. Double fertilization of ovules B. Give rise plant embryo and endosperm
C. Both A & B D. None

108. Which factor differentiate between plant development and animal development?
A. Plant cells do not migrate
B. Plant development is continuous
C. Plants have tremendous developmental plasticity
D. All of these

109. Primary metabolite consists of :
A. Amino acid B. Sugar
C. Nucleotides D. All of these

110. Which is not directly involved in the normal growth and development, or reproduction of an organism?
A. Primary metabolite B. Secondary metabolite
C. Both A & B D. None

111. Secondary metabolite derived from:
A. Primary metabolite B. Protein
C. Both A & B D. Streptocin

112. The totipotency of plant cells was predicted in 1902 by:
A. Haberlandt B. Hanning
C. Kotte D. None

113. Plant regeneration from cultured cells is proving to be a rich source of genetic variability is called
A. Somaclonal variation B. Germinal variation
C. Base variation D. All of these

114. Which technique is known for quantitative improvement of cell in ivitro?

A. Callus culture
B. Endosperm culture
C. Ovary/ Ovule culture
D. None

115. The meristematic tissues are differentiated into simple or complex tissues called :

A. Differentiation
B. Dedifferentiation
C. Redifferentiation
D. All of these

116. Reversion of mature tissues into meristematic state leading to the formation of callus is called:

A. Dedifferentiation
B. Dedifferentiation
C. Redifferentiation
D. None

117. The ability of callus to develop into shoots or roots or embryoid is called:

A. Redifferentiation
B. Dedifferentiation
C. Both A & B
D. Plasticity

118. The inherent potentiality of a plant cell to give rise to entire plant and its capacity is often retained even after the cell has undergone final differentiation in the plant system is described as cellular :

A. Totipotency
B. Gene
C. Both A & B
D. None

119. Spontaneous production of haploids usually occurs of the following process :

A. Parthenogenesis
B. Sexual reproduction
C. Monoceous
D. Semygamy

120. Shikonin is a secondary metabolite extracted from:

A. Root of *Lithospermum erythrorhizon*
B. Root of *Panax ginseng*
C. Both A & B
D. None

121. Plant possess a gametophytic number of chromosomes in their sporophytes called:

A. Haploid
B. Mutant
C. True diploid
D. None

122. Which mode is responsible for the development of haploids in nature?

A. Apomixis
B. Parthenogenesis
C. Both A & B
D. None

123. In 1964, Guha and Maheshwari observed the direct development of embryos and plantlets from microspores of :

A. *Daruta innoxia*
B. *A. indica*
C. *Hordeum vulgare*
D. All of these

124. Who obtained complete haploid plant in *Nicotiana tabacum* in 1967?

A. Guha and Maheshwari
B. Guha
C. Bourgin and Nitsch
D. Webber

125. Which approach is useful to determine the ploidy level of an organism?
A. Counting of plastid in the plant stomata
B. Chromosome number counting
C. Flow cytrometric analysis
D. All of these

126. Which is the quickest method to find out the status of ploidy level in plant:
A. Flow cytometry analysis
B. Number of nucleoli
C. Number of chromosome
D. None

127. Which is useful in diploidization?
A. Colchichine treatment
B. Endo-mitosis
C. Both A & B
D. None

128. Haploid is used in:
A. Hybrid development
B. Induction of mutation
C. Developing asexual lines of tree
D. All of these

129. Which variety is released in wheat through anther culture?
A. Florin and Zing hua 3
B. Hua 9 and Nonhua 5
C. Tanyu 2 and Tanyu 3
D. None

130. Who coined the term protoplast in 1880?
A. Hanstein
B. Klercker
C. Cocking
D. Takebe

131. Who described the enzymatic approach for isolation of protoplast from higher plant in 1960?
A. Cocking
B. Hanstein
C. Tabeke
D. None

132. Cell without cell wall is known as:
A. Cell wall
B. Protoplast
C. Cytoplasm
D. Nucleoplasm

133. Which enzyme is used for isolation of protoplast from cell
A. Cellulase
B. Hemicellulase
C. Pectinase
D. All of these

134. Which enzyme is used to digest the cellulosic component in the cell wall?
A. Pectinase
B. Hemicellulase
C. Acetyl Co A transferase
D. None

135. Which enzyme is responsible for degradation of cellulosic components of cell wall?
A. Cellulase
B. Hemicellulase
C. Both A & B
D. None

136. Which enzyme is used for isolation of protoplast from gram positive bacteria?
   A. Lysozyme
   B. Chitinase
   C. Xylanase
   D. All of these

137. Unit of biology which is composed of a cell's nucleus and the surrounding protoplasmic materials is known as:
   A. Protoplasm
   B. Cytoplasm
   C. Nucleus
   D. None

138. Which component used for maintaining the osmolarity in culture?
   A. Sorbitol
   B. Mannitol
   C. Both A & B
   D. Sugar

139. Which component concentration is required for more stability of membrance?
   A. Calcium
   B. Sorbitol
   C. Mannitol
   D. None

140. Nuclear genome come to hybrid and plastome from both parents is called :
   A. Cybrid
   B. Vybrid
   C. Both A & B
   D. Protoplasm fusion

141. Somatic hybridization involved the cross between:
   A. Protoplast
   B. Hybrid with inbred
   C. Both A & B
   D. Embryogenesis

142. Protoplast fusion consists of the following phase:
   A. Agglutination or adhesion
   B. Plasma membrane fusion at localized site
   C. Formation of heterokaryon
   D. All of these

143. A cell with two separate nuclei formed by the experimental fusion of two genetically different cells is called :
   A. Heterokaryon
   B. Homokaryon
   C. Homologous
   D. None

144. The term *heterokaryon* was coined in:
   A. 1965
   B. 1966
   C. 1987
   D. None

145. A cell that has multiple, identical nuclei in common cytoplasm is called :
   A. Homokaryon
   B. Heterokaryon
   C. Heterologous
   D. None

146. Which is not true in respect of somatic hybridization?
   A. Overcomes the sexual barriers
   B. Produced only hybrid
   C. Usable for gene transfers
   D. Usable to alteration of chromosome of an organism

147. Which of the following limitation found in somatic hybridization?
   A. End product of somatic hybridization is unbalanced
   B. Product of somatic hybridization has poor genetic stability
   C. Lack of efficient selection method of somatic hybridization
   D. All of these

148. Which of the following is essential component of breeding program designed to improve the traits of crops plants?
   A. Genetic variability   B. Mutation
   C. Selection   D. None

149. Who proposed the term somaclonal variation in 1981?
   A. Larkin   B. Scowcroft
   C. Both A & B   D. Mather

150. Variation arising due to culture of cell is known as:
   A. Somaclonal variation   B. Epigenetic variation
   C. Phytohormonal variation   D. Environmental variation

151. Genetic heterogeneity in culture arise due to:
   A. Mosaicism in chromosome   B. Irregularities in culture via natural mutation
   C. Error in mitosis   D. All of these

152. Calliclones indicates to:
   A. Variation arise in regenerated plants from stem
   B. Variation arise in regenerated plants from protoplast
   C. Both A & B
   D. Variation arise in regenerated plants from callus

153. Which way used for isolation of somaclonal variants?
   A. Without in vitro selection   B. With in vitro selection
   C. Both A & B   D. None

154. Which factor influenced the somaclonal variation?
   A. Genotypes and explants source
   B. Duration of cell culture and condition of culture
   C. Both A & B
   D. None

155. Improved variety DAMA developed via pollen haploid somaclonal method in which of the crop?
   A. Rice   B. Maize
   C. Castor   D. Brassica

156. Which is not correct?

A. Bio 13- *Citronella*
B. Pusa Jai Kisan- *Brassica*
C. Hasuyume- *Oriza sativa*
D. Varuna- *Brassica*

157. Who was first tested the possibility of in vitro selection for disease resistance against wild fire disease of tobacco in 1973?

A. Carlson
B. Larkin
C. Max
D. None

158. Freezing tolerance in Norstar winter wheat developed by:

A. Lazar *et al.* (1988)
B. Dorffling (1997)
C. Larkin (1981)
D. None

159. Which of the following is basis of somaclonal variation?

A. Karyotype changes
B. Cryptic chromosome rearrangement
C. Cytoplasmic genetic changes
D. All of these

160. Which of the following is major aim based on germplasm conservation?

A. To ensure the availability of useful germplasm at any time
B. Maintain the variability within the germplasm to long time
C. Both A & B
D. None

161. Preservation in the frozen condition is known as:

A. Cryopreservation
B. *Invitro* preservation
C. Both A & B
D. None

162. Which is not cryoprotectant?

A. Glycerol and dimethyl sulfoxide
B. Glycols
C. Acetamides
D. None

163. Which of the following is excellent cryoprotectant?

A. DMSO
B. Glycols
C. Proline
D. Mannose

164. Which concentration of DMSO is to be used for cryopreservation?

A. 5-10 %
B. 10-20%
C. 20-30%
D. None

165. Which is not true in respect of somatic hybridization?

A. Somatic hybridization is capable to cell fusion between incompatible parents
B. Somatic hybridization overcome the conventional breeding barriers
C. Somatic hybridization is not used for major or minor gene transfer
D. None

166. Which of the following enzyme is strongly controlled coiling of DNA helix?

A. Gyrase B. Topoisomerase

C. Both A & B D. Linkage

167. The genetic information flow in which of the following direction?

A. Unidirectional B. Bidirectional

C. Colinear D. All of these

168. Gene is defined as:

A. A segment of genome with a specific sequence of nucleotides

B. Unit of inheritance

C. Both A & B

D. None

169. Renaturation rate and DNA sequence complexity define which of the following curves?

A. Cot curve B. J-curve

C. Both A & B D. Exponential curve

170. The C_0t value is the product of:

A. The initial concentration of DNA and t (time in seconds)

B. DNA reassociation kinetics

C. Both A & B

D. None

171. Which statement is not related to C_0t analysis?

A. Usable for find out the repetitive DNA in simple DNA

B. Based on the principles of DNA reassociation kinetics

C. Both A & B

D. None

172. Which of the following RNAs are translated direct copy of the gene?

A. mRNA B. rRNA

C. tRNA D. snoRNA

173. Which of the following is end product of gene expression and act upon their roles in cell as RNA molecules?

A. rRNA B. tRNA

C. Both A & B D. None

174. Which of the following gene introduced into *Brassica napus* by Shewmaker in 1999 and enhanced the content of carotenoides?

A. Phytoene synthase gene B. Phytase gene

C. Both A & B D. S-adenosyl Methionine decarboxylase

175. The Flavr Savr is a variety of tomato released commercially by competent authority in:
A. 1665
B. 1965
C. 1976
D. 1988

176. Which method is used for Bt trait movement from one cultivar to another?
A. Mass selection
B. SSD
C. Introduction
D. Backcrossing method

177. Which of the following is not correct point in view of resistance management strategy against biotic factor for transgenic crops?
A. Reduction of selection pressure
B. Monitoring of populations
C. Alternate mortality factors
D. All of these

178. Manipulating nucleic acid of an organism is called:
A. Genetic engineering
B. Botany
C. Recombinant DNA analysis
D. None

179. Genes are transcribed into
A. mRNA
B. Enzyme
C. Visible traits
D. Protein

180. Bt cotton is ineffective against many cotton pests such as:
A. Plant bugs,
B. Stink bugs, and
C. Aphids
D. All of these

181. A combination of DNA molecules that are not found together in nature, produced by joining DNA obtained from different biological sources is called:
A. Recombinant DNA
B. zDNA
C. tRNA
D. Recombination

182. Which of the following amino acid is known as non- proteinaceous but play a key role in many physiological process?
A. β-alanine
B. Histamine
C. Serotonin
D. All of these

183. Which reaction is catalyses by lyases classes of enzyme?
A. Group elimination to form double bond
B. Bond formation coupled with ATP hydrolysis
C. Hydrolysis
D. None

184. Maximum enzymes in a cell are present in:
A. Mitochondria
B. Nucleus
C. Centrioles
D. None

185. tRNA bind which site during protein biosynthesis:

A. P site
B. A site
C. E site
D. R site

186. Strategy and methodology of Human genome project includes which of the following?

A. Only mapping
B. Only sequencing
C. Only generation of physical and genetic maps
D. All of these

187. Which of the following is known as first molecule of heredity?

A. RNA
B. DNA
C. Both A & B
D. None

188. A small RNA fragments is known as RNA primer has how much base pairs approximately:

A. 10-16
B. 20-30
C. 30-60
D. 40-50

189. Which of the following protein is prevented the recoiling of DNA?

A. SSBP
B. SSR
C. SNP
D. None

190. Who was first genetic surgeon?

A. Gregor Johann Mendel
B. Louise Brown
C. Galton
D. None

Terms 14

- **Batch culture:** Cell suspension grown in fixed volume of liquid medium.
- **Biochemical marker:** A marker includes protein based markers ie., isozymes
- **Callus:** A tissue arising from disorganized proliferation of cells either in culture or in nature.
- **cDNA:** A form of DNA artificially synthesized from a messenger RNA template and used in genetic engineering to produce gene clones. Complementary DNA is the one obtained from the reverse transcription process of RNA.
- **Chromosome walking:** A sequential isolation of clones carrying overlapping sequences of DNA, allowing large regions of chromosome to be covered, usually, to reach a locus of interest.
- **Clonal propagation:** Asexual multiplication starting from single individual.

- **Co-dominant marker:** An idea of marker as co-dominant; a condition refers to absence of intra locus interaction. Co-dominant is helped to identification of heterozygote into homozygotes.
- **Cryopreservation:** Preservation of plant propagative materials for long periods of time in frozen state at –196°C in liquid nitrogen.
- **Cybrids:** Somatic hybrids with normal protoplast of one species and such protoplast of another species, which either have no nucleus or have an inactive nucleus.
- **Cytological marker:** Markers that are related to alteration in chromosome number, shape and size.
- **Dedifferentication:** It refers to the reversion of a differentiated cell to the meristematic state.
- **Differentiation:** The modification of identical cells derived from a zygote into the different cell type.
- **DNA fingerprinting:** A technique used especially for identification by extracting and identifying the base-pair pattern of an individual's DNA.
- **DNA hybridization:** It is the process of combining two samples of denatured DNA.
- **Explant:** A piece of tissue used to initiate tissue culture.
- **Gene Therapy:** Gene therapy is defined as a set of strategies that modify the expression of an individual's genes or that correct abnormal genes. Each strategy involves the administration of a specific DNA or RNA.
- **Genetic engineering:** Manipulation of genetic architecture of an organism of molecular level.
- **Genetic map:** A list of ordered genetic loci for a particular genome.
- **Haplotype:** A set of linked alleles. Often this will refer to DNA sequence variation within a gene where one sequence variant may occur in the presence or absence of another.
- **Heterokaryon:** A cell in which two or more nuclei of unlike genetic makeup are present.
- **Heterologous probe:** A probe that is similar to, but not exactly complementary to the nucleic acid sequence.
- **Homokaryon:** A cell with two or more nuclei of similar genetic makeup.
- **Homologous probe:** A probe that is exactly complementary to the nucleic acid sequence.
- **Map unit:** A map unit is that distance in a chromosome, which permits one per cent recombination between two linked genes.
- **Marker assisted selection:** Selection of desirable traits with the help of DNA markers.
- **Marker:** Any segregating character or markable property whose pattern of inheritance can be followed at different levels.
- **Micro-propagation:** Production of miniature planting material by vegetative multiplication through growth hormone regeneration.
- **Molecular breeding:** The enhancement the yield and its related traits in plant or crops by using molecular engineering like using molecular marker or biotechnology.

- **Molecular marker:** A DNA sequence which is readily detected and segregates.
- **Morphological marker:** Any visual traits generally corresponds qualitative traits.
- **Orthology:** Literally 'the same word'. Genes in two different species are said to be orthologous if they are descended from the same single gene in their most recent common ancestor.
- **Parthenogenesis:** Embryos develop from the embryo sac without fertilization.
- **Physical map:** In physical map, distance between genes is shown in base pairs.
- **Probe:** A small nucleotide sequence that is complementary to a specific gene or nucleic acid sequence of interest.
- **Protoplast:** Cell without cell wall.
- **Recombinant DNA:** A DNA molecule produced by joining DNA segments from two different sources.
- **Recombination frequency:** Total number of recombinants divided by the total number of progenies in a test cross.
- **Redifferentiation:** The ability of callus to develop into shoots or roots or embryoid.
- **Restriction map:** A straight line representing a DNA segment of defined length.
- **Restriction mapping:** A technique is used for construction of restriction map of define DNA segment.
- **Retrotransposon:** A transposable element that undergoes transposition from one place to another in the genome of a cell by forming an intermediate RNA transcript from which a copy of the DNA of the transposable element is made using a reverse transcriptase and inserted into the genome at a new location is called retrotransposon.
- **RFLP (restriction fragment length polymorphism):** Variation in the length of a DNA fragment produced by a specific restriction enzyme acting on DNA from different individuals that usually results from a genetic mutation and that may be used as a genetic marker.
- **Secondary metabolite:** Chemical substances, which are found in cells but are not necessary for their survival.
- **Somaclonal variation:** The heritable variation for characters present in the cells cultured *in vitro*.
- **Somatic hybridization:** Hybrids developed by fusion of somatic cells.
- **Suspension culture:** Culturing of cells or cell aggregates in liquid medium.
- **Synkaryon:** Hybrid cell produced by fusion of nuclei in heterokaryon.
- **Totipotency:** The potential or inherent capacity of a plant cell or tissue to develop into an entire plant if suitably stimulated.
- **Transduction:** Introduction of foreign genetic material in to cell by means of a virus.

ANSWERS

Chapter 14: Molecular marker and Biotechnology

1. A	36. D	71. A	106. C	141. A	176. D
2. A	37. D	72. D	107. C	142. D	177. D
3. A	38. A	73. B	108. D	143. A	178. A
4. D	39. A	74. C	109. D	144. A	179. A
5. C	40. A	75. D	110. B	145. A	180. D
6. A	41. A	76. C	111. A	146. B	181. A
7. B	42. D	77. C	112. A	147. D	182. D
8. B	43. A	78. C	113. A	148. A	183. A
9. D	44. A	79. A	114. A	149. C	184. A
10. D	45. A	80. A	115. A	150. A	185. B
11. C	46. C	81. A	116. A	151. D	186. D
12. D	47. D	82. A	117. A	152. A	187. A
13. D	48. A	83. A	118. A	153. C	188. A
14. A	49. A	84. A	119. A	154. C	189. A
15. A	50. C	85. B	120. A	155. A	190. A
16. A	51. D	86. B	121. A	156. D	
17. A	52. C	87. A	122. C	157. A	
18. C	53. D	88. A	123. A	158. A	
19. A	54. A	89. C	124. C	159. D	
20. A	55. A	90. D	125. D	160. A	
21. C	56. C	91. A	126. A	161. C	
22. A	57. D	92. A	127. C	162. D	
23. D	58. B	93. A	128. D	163. A	
24. A	59. A	94. A	129. A	164. A	
25. D	60. A	95. A	130. A	165. C	
26. D	61. B	96. D	131. A	166. C	
27. A	62. A	97. D	132. B	167. B	
28. A	63. C	98. A	133. D	168. C	
29. C	64. A	99. C	134. D	169. A	
30. A	65. A	100. C	135. A	170. A	
31. B	66. A	101. C	136. A	171. C	
32. D	67. A	102. D	137. A	172. A	
33. B	68. A	103. A	138. C	173. C	
34. C	69. A	104. D	139. A	174. A	
35. A	70. C	105. A	140. A	175. B	

Chapter **15**

Gene Transfer in Plant and Vector

1. Transfer of gene by passing sexual process is known as:
   A. Transgene B. DNA
   C. Transgenic D. Vector
2. An organism developed by genetic engineering is known as:
   A. Transgenic B. Host plant
   C. Donor plant D. Transgenes
3. The condition arises when transferred gene or DNA may be expressed for short period:
   A. Transient expression B. Dominant expression
   C. Codominant expression D. Epitasis expression
4. The term cybrids coined by :
   A. Scow Croft B. Larkin
   C. Kumar *et al.* D. Bunn *et al.*
5. Marker genes are categorized in the following types:
   A. Reporter gene marker B. Selectable gene marker
   C. Both A & B D. None
6. Marker gene is also known as:
   A. Genetic marker B. DNA sequence
   C. Both A & B D. Factor
7. Which of the following feature is not present in reporter gene.
   A. Detection with high sensitivity B. High endogenous background
   C. Assay should be nondestructive D. All of these

8. Which of the following synthase genes found in T DNA of Ti or Ri plasmids of *Agrobacterium*?
   A. Octopine synthase genes
   B. Nopaline synthase genes
   C. Both A & B
   D. None
9. Who discovered restriction enzyme in 1962?
   A. W. Arber
   B. D. Nathans
   C. H. Smith
   D. EM Southern
10. Which of the following step is used for molecular cloning?
   A. Choice of host organism, cloning vector, and preparation of vector DNA
   B. Preparation of DNA to be cloned, creation of recombinant DNA and introduction of recombinant DNA into host organism
   C. Selection of organisms containing recombinant DNA and screening for clones with desired DNA inserts and biological properties
   D. All of these
11. Which enzyme is involved in creation of recombinant DNA?
   A. Ligase
   B. DNA polymerase
   C. Primase
   D. All of these
12. Which technique is more efficient for gene transfer in bacterial, animals and plant cells?
   A. Lipofection
   B. Conjugation
   C. Transformation
   D. All of these
13. Liposomes are :
   A. Circular lipid molecule
   B. Linear lipid molecule
   C. Collinear lipid molecule
   D. None
14. A natural microbial recombination process by which single stranded DNA is transferred via cytoplasmic bridges from the donor bacteria to the recipient bacteria is known as:
   A. Conjugation
   B. Transduction
   C. Lipofection
   D. None
15. Which of the following technique is not suitable for direct gene transfer?
   A. Chemical mediated gene transfer
   B. Particle gun/Particle bombardment
   C. Lipofection
   D. None
16. Which of the following method of gene transfer does not require vector?
   A. Particle gun/particle bombardment method and lipofection method
   B. Electroporation and Microinjection techniques
   C. Transformation and conjugation techniques
   D. All of these
17. Which is known as "natural genetic engineer" of plants?
   A. *Agrobacterium tumefaciens*
   B. E. coli
   C. Neurospora
   D. Drosophilla

18. Which is true in respect of particle gun/particle bombardment method?
    A. Two types of plant tissue are commonly used for particle bombardment- Primary explants
    B. The $CaCl_2$ breaks the cell wall at certain regions and binds the DNA to the cell surface.
    C. It involves a pulse of high voltage applied to protoplasts/cells/ tissues to make transient (temporary) pores in the plasma membrane which facilitates the uptake of foreign DNA.
    D. None
19. Which of the following technique is used to study the segregation and stability integration and expression of transgene in the subsequent generations?
    A. PCR technique B. Western blot
    C. Southern and Northern technique D. All of these
20. A *HpaI* is an enzyme encoded from:
    A. *Hemophillus parainfluenzae* B. *H. influenzae*
    C. *H. aegyptius* D. All of these
21. Which is not correct?
    A. Hemophillus influenza - Hinc II and Hind III
    B. Himophillus parainfluenzae –Hpa I and Hpa II
    C. Bacillus amyloqueffaciens –Bam HI
    D. E.coli - Msp I
22. Who got Noble Prize in 1978 for medicine?
    A. H Smith B. D Nathans
    C. W Arber D. All of these
23. Which technique is used for separation of large DNA molecules?
    A. Pulsed field gel electrophoresis B. Dote blots and slot blot techniques
    C. Lipofection technique D. None
24. Time unit is equal to:
    A. One minute B. One second
    C. One hour D. One office hour
25. Who described Pulsed field gel electrophoresis technique?
    A. Schwartz and Cantor B. E.M. Shourthen
    C. Alec Jeffreys D. Paul Berg
26. Which of the following technique is better than lipofection technique?
    A. Chemical phosphate method B. Transformation transfection technique
    C. Conjugation technique D. Transduction technique
27. Who described the Phospholipids as Gene Delivery Vehicles in 1980?
    A. Schaffner B. Szybalska and Szybalska
    C. Van der Erb D. Orrantia and Chang

28. Which of the following method is low toxic transfection method?
   A. Chemical transfection method
   B. Lipofection
   C. Both A & B
   D. None
29. Who introduced successfully plasmid DNA into mouse fibroblasts in 1982?
   A. Wong and Neumann
   B. Szybalska and Szybalska
   C. Van der Erb
   D. Orrantia and Chang
30. To seal a DNA fragment into a bacterial plasmid, which of the following enzyme is essential?
   A. Ligase
   B. Primase
   C. Transporases
   D. Helicase
31. Which of the following is not required in PCR?
   A. Taq
   B. Primer
   C. Helicases
   D. Annealing temperature
32. Which of the following is responsible for separation of complementary strand of DNA?
   A. DNA polymerase
   B. High pH
   C. RNA polymerase
   D. All of these
33. Which process exhibits the separation of complementary strand of DNA into single strand of DNA?
   A. Denaturation
   B. Renaturation
   C. Conversion of DNA into mRNA
   D. DNA synthesis
34. Renaturation rate refers to:
   A. Ratio between time taken for renatured DNA and denatured DNA
   B. No. of DNA helix
   C. Proportional of DNA double strand in biosynthesis of protein
   D. None
35. Nucleic acid hybridization techniques include which of the following?
   A. DNA-DNA hybridization
   B. DNA-RNA hybridization
   C. Both A & B
   D. Base –Base covalent techniques
36. Which is not true in respect of southern blot?
   A. Use to detect methylated sites in particular genes
   B. Use of homology-based cloning on the basis of amino acid sequence of the target gene
   C. Use of detection of a specific DNA sequence in DNA samples
   D. None
37. Which of the followings, feature is not present in a vector?
   A. dsDNA and ring shape
   B. Linearity and take donor DNA fragment
   C. Only self replication
   D. Autonomous element

38. Based on usefulness, type of vector is not categorized in which of the following?
A. Cloning vector B. Suttle vector
C. Expression vector D. None

39. Which of the following is not a property of plasmid/vector?
A. Autonomous element B. Linearity
C. Ring replication D. Easy multiplication

40. The inserted DNA is also known as:
A. Polylinker DNA B. Promiscuous DNA
C. Donor DNA D. cpDNA

41. Which of the following DNA is mostly used as a vector?
A. Circular DNA B. Promiscuous DNA
C. Linear DNA D. Double stranded DNA

42. Who used the term plasmid for extrachromosomal hereditary determinant in 1952?
A. Mc Clintock B. H. Wallace
C. M.L. Birnstiel D. J. Lederberg

43. Which of the following is not true in respect of episomes?
A. It is a RNA fragment B. Autonomous element
C. Originate via conjugation D. None

44. Extranuclear accessory genetic element is known as:
A. Vector B. Plasmid
C. Cytoplasm D. Both A & B

45. Which of the following genetic material is known as episomes?
A. *Col* factor B. Bacteriophage
C. Sex factor D. All of these

46. Which is not true in respect of episomes?
A. All episomes are known as accessory DNA molecules.
B. It is also known as IS element.
C. Extra-nuclear accessory genetic determinant is known as episomes
D. It is also known as plasmid.

47. Which of the following, strain was used for development of R plasmid/R Factor in 1956 at Japan?
A. Shigella B. *Salmonella*
C. *Escherichia* D. All of these

48. Who discovered *Shigella* in 1897?
A. *Kiyoshi Shiga* B. Bridges
C. Tatum D. None

49. Colicins is a toxin, which is produced by which of the following bacteria?
    A. *Escherichia*　　B. *Shigella*
    C. *Salmonella*　　D. All of these
50. Which of the following bacteria is not gram negative?
    A. *Shigella*　　B. *Escherichia*
    C. *Bacillus*　　D. None
51. Which of the following is resembled to col factor?
    A. F-factor　　B. R-Factor
    C. Both A & B　　D. None
52. Which of the following is responsible for production of colicin?
    A. R factor　　B. Sex factor
    C. Both A & B　　D. Col factor
53. Which factor is responsible to transfer of genetic material from one strain to another?
    A. Sex factor　　B. Conjugation
    C. R factor　　D. Transduction
54. The fertility trait is commonly found in which of the following?
    A. F-plasmid　　B. R6 plasmid
    C. Col　　D. Both A & C
55. Which of the following plasmid is responsible for creation of tumorigenicity in plant?
    A. ColEi plasmid　　B. R plasmid
    C. F plasmid　　D. None
56. DNA plasmid represents, which type of replication?
    A. Semi-conservative　　B. Dispersive
    C. Conservative　　D. None
57. In *E.coli,* transfer of plasmid is controlled by which of the following?
    A. Replication　　B. Mutation
    C. Transcription　　D. None
58. Bacterial chromosome are insertion sequence which are homologue in which of the following?
    A. Bacteria and vector　　B. Bacteria and eukaryotic DNA
    C. Plasmid and bacteria　　D. Plasmid and animal kingdom
59. The segment of DNA which can move from one location to another is known as:
    A. Promiscuous DNA　　B. Mobile element
    C. Insertion element　　D. None
60. The certain sequences are capable for moving from one site to another is known as :
    A. Mobile element　　B. Transposable element
    C. Transposons　　D. All of them

61. Which of the following transpon reported?

A. IS element
B. P element
C. Ac/Ds
D. None

62. IS element was detected first of all in which of the following?

A. Gal operon in *E. coli*
B. Neurospora
C. Both A & B
D. None

63. Which point does not match with pBR 322?

A. Widely used in gene cloning.
B. Derived from ColEI and have 4362 bp
C. Contain resistance against two antibiotic
D. Carry HincII fragment of lac Z

64. Which is best vector for transferring the gene in plant?

A. Ti plasmid
B. YAC
C. pBR322
D. pUC

65. Which DNA is found in bacteriophage?

A. Ring DNA
B. Collinear DNA
C. Linear DNA
D. Haploid DNA

66. Which of the following plays vital role in recombinant DNA technology:

A. Ligase
B. Restriction enzyme
C. Both A & B
D. Vector

67. Which of the following is natural occurring fragment?

A. Plasmid
B. Chimeric DNA
C. Both A & B
D. None

68. A replicating DNA molecule is known as:

A. Cloning vector
B. Virus
C. Bacteria
D. None

69. Which is true in relation to ëgt10 ?

A. 7kb in length
B. 6 kb in length
C. 43.7 kb dsλ phage for cloning DNA fragment
D. None

70. Which of the following is used in gene cloning?

A. DNA Ligase
B. DNA Gyrase
C. DNA helicase
D. DNA polymerase

71. pBR 322 is encoded by which of the following?
    A. Resistance gene against chlorapmphenicol
    B. Resistant gene against ampicillin
    C. Both A & B
    D. Resistant gene against kanamycin
72. Which is the first artificial cloning vector?:
    A. pUC B. pBR 322
    C. pACY 184 D. None
73. Plasmid is a:
    A. Closely circular DNA molecule B. Inheritance determinant
    C. Like behave as accessory genetic unit D. All of these
74. Virulence plasmid encodes, which of the following?
    A. Drugs B. Toxin
    C. Conjugate gene D. None
75. Plasmid size varies between:
    A. 1 to 200 kb B. 500-1000 kb
    C. 300-400 kb D. None
76. Synthetic cloning site is also known as:
    A. Polylinker B. Restriction site bank
    C. Polycloning site D. All of these
77. Histochemical test is used for:
    A. Visual identification of recombinant clone
    B. To identification of allele in plant
    C. To find out the tissue of cell
    D. None
78. The gene determining tetracycline resistance was derived from:
    A. pSC101 B. pUC
    C. pMB1 D. p15A
79. How many restriction sites are found within the pBR 322?
    A. 20 B. 2
    C. 12 D. 23
80. Which of the following microbe are unicellular eukaryotes?
    A. *Sacchromyces cerevisiae* B. *Escherichia coli*
    C. *Neurospora spp.* D. None
81. Which one is different from others of the following?
    A. Yeast integrating plasmid B. Yeast episomal plasmid
    C. Yeast replicating plasmid D. Yeast centromeric plasmid

82. Which sentence is not true in respect of Ylps?

A. Yeast integrating plasmid is not actually yeast plasmid.

B. Integration is essential for successful transformation.

C. Integration is due to non-homologous recombination between the yeast gene and the active gene in yeast chromosome.

D. Lack autonomous replication.

83. Yeast replicating plasmid differs from yeast integrating plasmid because:

A. Yeast replicating plasmid has autonomous replicating property.

B. It is based on chromosomal element.

C. Highly frequent to transformation.

D. All of these

84. ARS centromeric plasmid is referred to as:

A. Mini-chromosome B. Linear chromosome

C. Colinear chromosome D. None

85. Yeast linear plasmid is based on:

A. ARS centromeric plasmid B. Linear rather than circular

C. It may be replicate in *Escherichia coli.* D. All of these

86. Which is not true for linear plasmid?

A. Developed by adding telomeric sequence to the end of a cut ARS based plasmid

B. Linear plasmid are present like a guanine chromosome, single copy

C. More stable yeast plasmid vector

D. Linear plasmid can also be replicate in *E.coli.*

87. Which disease is caused by a bacteria that carries Ri plasmid?

A. Hairy root disease B. Crown gall disease

C. Foolish seedling disease D. Leaf and color disease

88. Which disease is caused by a bacteria that codes Ti plasmid?

A. Hairy root disease B. Crown gall disease

C. Foolish seedling disease D. Leaf and color disease

89. Which of the following is a basic component of bacteriophage?

A. Protein B. Nucleic acid

C. Both A & B D. None

90. Which of the following term is defined as capacity for transfer foreign DNA from phage genome to bacterial host?

A. Transfection B. Transformation

C. Conjugation D. None

91. Lambda DNA replication proceed to bidirectional and formed which of the following structure?

A. Theta
B. Omega
C. Sigma
D. Beta

92. Which of the following gene inhibits the transcription in gene cloning:

A. λcl gene
B. Z
C. Both A & B
D. None

93. Which of the following does not present cosmid formation?

A. Hybrid plasmid that contains a Lambda phage cos sequence
B. P1 and F Factor bind together as formed cosmid
C. Both A & B
D. None

94. Which of the following is the length of λ phage?

A. 48503 bps
B. 200 bps
C. 1000 bps
D. None

95. Which feature is not found in cosmid?

A. It possess origin, a selectable gene marker and suitable cloning site
B. It is a ideal vectors for genome mapping
C. It has not lytic gene
D. In vitro packaging is not needed to maintenance the cosmids in virues

96. The capacity of cosmids to take inserted DNA varies:

A. 40-45 kb
B. 50-55 kb
C. 60-70 kb
D. None

97. Lambda genome is a :

A. Double stranded DNA molecule
B. Single stranded DNA molecule
C. Both A & B
D. None

98. Which is not true in respect of lambda phage?

A. dsDNA has 48.5kb in length
B. DNA is linear with short complementary single stranded projections of 12 nucleotides at its 5 prime end
C. It is not capable to kill the bacteria
D. None

99. M13 phage is isolated from:

A. *E. coli*
B. Yeast
C. *Neurospora*
D. None

100. Who developed YACs in 1987?

A. David Burke *et al.* B. Joachin Messing

C. Ratzkin D. None

101. pBluescript II KS is a vector based on one of the following plasmid:

A. pUC plasmid B. pBR 322

C. YAC D. M13 phages

102. Which of the following vector has polycloning site?

A. Phagemids B. Cosmid

C. pUC 18 D. pUC 19

103. Modification of hnRNA occurs in :

A. Nucleus B. Cytoplasm

C. Both A & B D. Ribosome

104. Which steps is known as first step in complementary DNA libraries?

A. Isolation of mRNA

B. Synthesis of 1st and 2nd strand of cDNA

C. Cloning of cDNA and induction to host cells

D. Clone selection

105. The large molecular weight DNA can be checked by which of the following technique?

A. Pulsed field gel electrophoresis B. Proteinase K method

C. SDS method D. All of these

106. Which method is also known as dideoxy-mediated chain termination method?

A. Sanger-Coulson method B. Biotin streptavidin labeling system

C. Maxam-Gilbert method D. None

107. Sanger method is not required for which of the following DNA sequencing?

A. Single stranded template

B. Specific primer

C. Klwnow fragment of *E.coli* DNA polymerase

D. Simple chemical reagents

108. Who got Noble Prize in chemistry for his valuable work on PCR in 1993?

A. Kary Mullis B. Ritzer

C. Benner D. None

109. Which of the following feature of PCR is known as key feature of PCR?

A. Exponential nature of amplification

B. Based on semi-conservative type replication in both prokaryotes and eukaryotes

C. Carry DNA polymerase enzymatic activity

D. None

110. Taq polymerase is a :
   A. Thermo-stable enzyme
   B. Isolate form Thermus aquaticus
   C. Taq's optimum temperature for activity is 75–80°C
   D. All of these

111. Which of the following is a drawback of Taq polymerase?
   A. Lack of 3' to 5' exonuclease proof-reading-activity
   B. It is catalyzed the amplification reaction
   C. The template DNA is 1st denatured by heating at 94°C
   D. None

112. Which of the following enzyme is not involved in PCR technique?
   A. Taq DNA polymerase
   B. Primase
   C. DNA Helicase
   D. None

113. Which name is known as nick name of polymerase chain reaction?
   A. Public chain reaction
   B. People chain reaction
   C. Poly public chain reaction
   D. None

114. The Taq DNA polymerase acts in best way in PCR at :
   A. 72°C
   B. 80°C
   C. 90°C
   D. None

115. What temperature is used to be for denaturation in PCR that does not destroy any enzymatic activity?
   A. 72 °C
   B. 80 °C
   C. 90 °C
   D. 96 °C

116. Which of the following enzyme is used for amplification of DNA molecule?
   A. Primase
   B. DNA polymerase and
   C. DNA helicase
   D. All of these

117. In PCR, which is the key of primer extension reaction?
   A. Taq DNA polymerase
   B. Helicase
   C. RNA polymerase
   D. Primes

118. Taq DNA polymerase and E.coli DNA polymerase I have which of the following activity in PCR?
   A. 5' to 3' exonuclease activity
   B. 3' to 5' exonuclease proof-reading-activity
   C. Both A & B
   D. None

119. Which of the following enzyme does not helpful for incorporation of mismatch bases?
   A. Taq DNA polymerase
   B. Helicase
   C. RNA polymerase
   D. All of these

120. Inverse peoples chain reaction can be used to:

A. Amplification of unknown DNA sequences

B. Useful method of chromosome crawling to explore the chromosome sequence

C. Both A & B

D. None

121. PCR based marker can be used in mapping of sequences of:

A. RAPDs B. SSRs

C. AFLPs and SNP D. All of these

122. *Agrobacterium* mediated gene transfer is extremely used in :

A. Maize B. Rice

C. Wheat D. All of these

123. DNA mediated gene transfer was successfully used in :

A. Monocotyledons B. Di-cotyledons

C. Both A & B D. None

124. Which method consisted DNA mediated gene transfer?

A. PEG stimulated DNA uptake

B. Microinjection and macro-projection

C. Microprojectile, electroporation and liposomes

D. All of these

125. The first transgenic plant developed at USA in :

A. 1883 B. 1983

C. 1783 D. None

126. First time, which country permits for the multiplication of transgenics at commercial levels?

A. USA B. India

C. Japan D. China

127. Herbicide tolerant transgenics developed in the following:

A. Corn and cotton B. Soybean and potato

C. Tomato D. All of these

128. In which, virus coat protein gene is not transferred for development of transgenics?

A. Rice B. Tomato

C. Alfalfa D. Cowpea

Terms

- **Annealing:** A step of DNA denaturation.
- **ARS:** Autonomous replicating sequences.
- **Conjugation:** Transfer of desirable DNA to host via conjugation tube or Cytoplasmic bridges or conjugate bridges.
- **Cosmids:** Plasmid vector into which phage lambda cos sites have been inserted resulting the plasmid DNA can be packaged in vitro in the phage coat.
- **DNA amplification:** A process by which multiplication of a piece of DNA is done in a test. The most commonly used method is PCR. Polymerase chain reaction.
- **Endonuclease:** Enzyme used in RDT for cutting the DNA segment at mid point.
- **Epicotyl:** Upper portion of the axis of a plant embryo or seedling above the cotyledon.
- **Episome:** A plasmid which replicates within the cell and can integrate into the host bacterial DNA.
- **Exonuclease:** Enzyme used to cut the DNA at terminal portion.
- **Gene cloning:** A way by which synthesis of multiple copies of DNA from a particular DNA sequence used for gene transfer into host cells.
- **Genetic marker:** A sequence of DNA to be used to track or mark a particular location on particular chromosome.
- **GMO:** An organism that has been modified by genetic engineering is called a GMO (genetically modified organism)
- **Histochemical test:** Test to be used for identification of recombinant clone or allele.
- **Inverse PCR technique:** A technique in which amplification of those DNA sequences take place, which are away from the DNA primers and not flanked by the primers.
- **Ligation:** A process by which two DNA segment are sealed by enzyme ligase.
- **pBR 322:** A standard cloning vector.
- **Plasmid:** A self replicated dsDNA segment exist extra-chromosomal complement unit found in cell.
- **Primase:** An enzyme creates the RNA primer for DNA replication.
- **Primer:** A short sequence of RNA or DNA from which DNA replication can initiate.
- **Primosome:** A complex of proteins, a primase and a helicase which results in the initiation of the synthesis the synthesis of RNA primers on the lagging DNA strand during DNA replication.

- **Recombinant DNA:** The DNA developed by joining of two dissimilar DNA segments.
- **Recombinant joint:** A point at two DNA molecules recombined.
- **Reporter gene:** A coding unit whose product is easily assayed during genetic transformation.
- **Restriction map:** A linear array of sites on DNA cleaved by different restriction enzyme.
- **Restriction site:** A base sequence where the enzyme nuclease cut the DNA.
- **RNA phage:** A phage which carries the RNA as genetic material.
- **Robertsonian fusion:** The fusion of two acrocentric chromosomes.
- **Shotgun technique:** A technique by which cloning a large population of a small DNA fragment.
- **Shuttle vector:** It is vector like plasmid constructed in eukaryotes and prokaryotes cell.
- **SINE:** Short interspersed element found in eukaryotic genome.
- **Stable transformation:** A recipient cell should be representing their capacity of regeneration into fertile plant. In other words, host cell received foreign DNA segment representing stable performance.
- **Ti Plasmid:** A plasmid causes the tumor result crown gall disease arise in plant.
- **Transfection:** A capacity for transfer foreign DNA from phage genome to bacterial host.
- **Transformation:** A process by which foreign DNA inserted into host cell.
- **Transgenic:** Organism developed through genetic engineering.
- **Transient expression:** A condition arises when transferred gene or DNA may be expressed for short period of the DNA transfer.
- **Vector:** A way used in recombinant DNA technology. In other words, a plasmid, phage, virus or bacterium carrying foreign DNA.
- **Virulent phage:** A phage leads always lysis in host cells.

ANSWERS

Chapter 15: Gene transfer in plant and vector

1. A	31. C	61. A	91. A	121. D
2. A	32. B	62. A	92. A	122. D
3. A	33. A	63. D	93. A	123.
4. D	34. A	64. A	94. D	124. D
5. C	35. C	65. C	95. A	125. B
6. C	36. D	66. C	96. A	126. D
7. B	37. B	67. A	97. A	127. D
8. C	38. D	68. A	98. C	128. D
9. A	39. B	69. A	99. A	
10. D	40. A	70. A	100. A	
11. A	41. A	71. B	101. A	
12. A	42. D	72. B	102. A	
13. A	43. A	73. D	103. A	
14. A	44. B	74. B	104. A	
15. D	45. D	75. A	105. A	
16. D	46. D	76. D	106. A	
17. A	47. A	77. A	107. D	
18. A	48. A	78. A	108. A	
19. D	49. D	79. A	109. A	
20. A	50. D	80. A	110. D	
21. D	51. C	81. C	111. A	
22. D	52. D	82. D	112. C	
23. A	53. A	83. A	113. B	
24. A	54. D	84. D	114. A	
25. A	55. D	85. D	115. C	
26. A	56. A	86. A	116. D	
27. A	57. B	87. B	117. A	
28. B	58. C	88. C	118. A	
29. A	59. A	89. A	119. D	
30. A	60. D	90. A	120. C	

Question Based Points

DNA polymerase:

1. The in vitro synthesis of DNA was first accomplished by A. Kornberg & his co-workers in **1957**; who received Noble Prize in **1959.** For this works, they isolated an enzyme from ***E.coli*** (initially called DNA polymerase or Kornberg enzyme, now known as DNA polymerase I) that catalyses the covalent addition of nucleotide to pre-existing DNA chains.
2. The enzyme is active only in the presence of Mg^{++} ions & pre-existing DNA.
3. DNA polymerase I catalyse the formation of a phosphodiester bridge between the **3'-OH at the end of the primer DNA chain and 5'-phosphate of the incoming DNA (5′-3′ synthesis).**
4. Fourth polymerase enzyme; named and has been isolated from **calf thymus and rabbit bone marrow.**
5. DNA polymerase I is the major replicative enzyme.
   Poly A mutants of ***E.coli*** replicate their DNA at normal rates.
6. Major function of DNA polymerase I is **DNA repair.**
7. The 3′→ 5′ exonuclease activity catalyses the removal of nucleotides, one by one, from the **3′ end of polynucleotide chains.**
8. DNA polymerase I of E.coli also has 5′→ 3′ exonuclease activity.
9. **The 3′→ 5′** exonuclease activity of DNA polymerase carried out a critical **"proof reading" or "editing"** function *i.e.* **essential for the high degree of fidelity characteristics of DNA replications.** Only DNA polymerase δ possesses the 3′→ 5′ exonuclease proof-reading activity.
10. The 5′→ 3′ exonuclease activity of many prokaryotic DNA polymerase is also very important. It functions in the removal of segment of DNA damaged by **UV light irradiation.**
11. 5′→ 3′ exonuclease activity has not been found in any of the **eukaryotic DNA polymerases.**
12. Polymerases have an absolute requirement for a free 3'-OH so as to carry out **5′ → 3′ synthesis (also direction known as leading strand).**
13. **3′→ 5′ directions of DNA synthesis, called** lagging strands (proof reading).
14. RNA polymerase, a complex enzyme that catalyses the synthesis of RNA molecule from DNA templates.
15. DNA synthesis is initiated by **RNA primers.**

16. DNA synthesis is primed by short segments of RNA which are later removed by a **5′→ 3′ exonuclease** and replaced by DNA prior to covalent sealing by polynucleotide ligase.
17. **RNA** primers are excised by the 5′→ 3′ exonuclease activity of DNA polymerase I.
18. **The synthesis of RNA primers is catalysed by enzyme called primases.**
19. The *E. coli* primase is the product of **dnaG gene.**
20. **In RNA primers (prokaryotic)** are 10-60 nucleotides in length.
21. RNA primers are used to initiate DNA synthesis.
22. DNA replication is complex, carried out by a multi enzyme complex, often called replication apparatus or replisome.
23. **Unwinding protein** (DNA helicase) catalyse the unwinding of the DNA double helix.
24. The DNA single strand binding proteins (SSBPs) bind to DNA as tetramers.
25. **DNA gyrase** which catalyse the formation of **negative super coils in DNA;** are essential for replication & are believed to play a key role in the "unwinding process".
26. Primase activity requires the formation of a complex of primase & atleast six other proteins, this complex is called **primosome.**
27. The covalent extension of the primed DNA chains during chromosome replication in *E. coli* is carried out by DNA polymerase III. DNA polymerase III contains given different polypeptides.
28. The 5′→ 3′ polymerase & 5′→ 3′ exonuclease activity are both **present on the α polypeptide of DNA polymerase III.**
29. 3′ → 5′ (proof reading or editing) activity of polymerase III is present on the **E polypeptide.**
30. Supercoiling is involved in **recombination, gene expression & regulation of gene expression.**
31. DNA gyrase (*E. coli*) activity is inhibited by the drugs **novobiocin and nalidixic acid,** two potent inhibitor of DNA synthesis in bacteria.
32. DNA gyrase is required for **DNA replication.**

DNA Repair:

33. Damage to DNA is any change in DNA that introduces a deviation from the usual double helical structure. The various damages to DNA may be grouped into the following two types; (*i*) single base changes (*ii*) change in DNA structure.
34. Single base changes (DNA damage) are represented by **conversion of one base into another.**
35. DNA repair may be grouped into five general categories:–

    (*i*) Direct repair (*ii*) Excision repair

    (*iii*) Mis-match repair (*iv*) Tolerance repair/system

    (*v*) Retrieval system

36. Simple removal of the damage to the DNA is known as **Direct repair.**

37. Removal of the covalent bonds between the two 5 & two 6 carbons of the two thymine reduces involved in thymine dimers.

38. Thymine dimers are generally formed **due to UV radiation & they interfere with replication and transcription.**

Excision Repair:

39. Recognition and incision step, damaged/mis-match section of a strand is recognised by an **endonuclease;** this enzyme then cuts the affected strand on both sides of damage.

40. After the incision, **a 5′→ 3′ exonuclease** digests away the damaged/mis-match selection.

41. In *E.coli,* **excision is most likely due to the 5′→ 3′ exonuclease activity of DNA polymerase I.**

42. The excision and synthesis events are most likely performed by the **DNA polymerases I.**

43. The **Uvr ABC** repairs accounts for about 99% of the excision repair activity in ***E.coli.***

44. **Recombination-repair system/post replication repair or retrieval systems operate** when a structural distoration blocks replication at the damaged sites.

45. The recombination repair pathway involves **Ree BC & Rec F** pathways.

46. The **Lex A protein** acts as a repressor of its own gene, of the Rec A gene & of several operons of *E.coli.*

47. The Lex A protein represses its target genes by binding to a **20 bp stretch of DNA** called the **SOS box.**

Transcription:

48. Genes determine the amino acid sequences of polypeptide.

49. In 1958, Waston Crick proposed that the information present in DNA is transferred to DNA (via replication) & RNA and from RNA to protein (form of amino acid sequence).

50. Relation between DNA, RNA & protein are known as **"Central dogma of Biology".**

51. The production of RNA copies of a DNA template is known as **Transcription;** its catalysed by **RNA polymerase or transcriptase.**

52. RNA polymerase activity was first shown in 1960 by **Wiess in rat liver.**

53. The sequence of transcription units are written in such a manner that the transcription proceeds from **left to the right.**

54. The sequence on the left of the start point is called **upstream sequence,** denoted by – 1, – 2 etc.

55. The sequene on the right start point is termed as **downstream sequence,** denoted by + 1, + 2 etc.

56. The RNA transcripts obtained directly from transcription are called **primary transcripts.**

57. The up element is an **A-T rich** region located on upstream of the **– 35 sequence**. The upstreams interacts with the α sub-units of RNA polymerase, and is found in promoters of such gene that are highly expressed *e.g.* rRNA genes.

58. Down mutation in the – **35 sequence** reduce the rate of closed binary complex formation.

59. **Continue the synthesis of RNA using DNA as template;** the complete enzyme is known as **holoenzyme.**

60. Holoenzyme is ~465 KD in weight, ~90 × 95 × 160 Å in size consists (*i*) core enzyme, (*ii*) sigma factor.

61. Sigma factor is a single polypeptide encoded by **rpo D** that ensures the RNA polymerase holoenzyme binds stably **only at the promoters.**

62. The binding of the enzyme to any DNA segment is called **loose binding.**

63. Only 1/3 of the RNA polymerase molecules can remain as holoenzyme.

64. Sigma factors make critical contacts with promoters DNA in the – 35 & – 10 regions.

65. Almost all the core enzyme exists in **association with DNA.**

66. When a factor (sigma factor) associates with a core enzyme complex with DNA at non-specific site, the **holoenzyme quickly dissociates** from the **loose binding site.**

Tempelate recognition:

67. The tempelate recognition phase of transcription begins with **binding of RNA polymerases holoenzyme** to the promoter region of a transcription unit.

68. Holoenzyme binds to the duplex DNA of the promoter region forming **closed primary complex.**

69. Closed Binary Complex- A complex of protein with normal DNA duplex.

70. The complex of **melted DNA** with the enzyme is known as **open binary complex.**

71. The conversion of a closed binary complex into an open binary complex is irreversible at **strong promoters.**

Initiation:

72. Transcription initiation showed that the synthesis of 1st nucleotide bond in RNA transcript.

73. The nucleotides used for RNA synthesis are **riboside 5'-triphosphate,** which have a free 3' OH, **which is an absolute requirement for** RNA synthesis.

74. The RNA polymerase permits only the complementary riboside to align against any given nucleotide of the template DNA.

75. A transcription bubble represents that region of a **DNA duplex being transcribed** & length ranges between **12-14 bp.**

76. The time taken for the RNA polymerase to leave the promoter is called **promoter clearance time (1-2 seconds).**

77. RNA polymerase I: is located in the **nucleolus** & is responsible for the transcription of **rRNA.**

78. RNA polymerase II: located at **nucleoplasm,** having 20-40% activity and **produced mRNA.**

79. RNA polymerase III: found in **nucleoplasm** & transcribed tRNA and other small RNA genes. It is inhibited by high level of **α aminitin in animal cell.**

80. The largest sub-unit of RNA polymerase II of yeast is homologous to *E. coli* **β' sub-unit** and binds DNA.

81. Second largest unit of RNA polymerase II is homologous to the **β' sub-unit of *E.coli* & binds nucleotides.**

82. The largest sub-unit of RNA polymerase II has a **Carboxy Terminal Domain (CTD),** which consists of multiple repeats of a consensus sequence of 7 amino acids.

83. **RNA polymerase I has not been sensitivity to α-amanitin.**

Transcription factors:

84. Basal transcription factors are required for initiation of transcription at all the promoters. Factors- TFII A, TFII B, TFII D, TFII E, TFII F, TFII H & TFII J (total 7 factors).

85. TBP (Tata Binding Protein) is a small protein of ~38 KD.

86. TBP play a same role for all the three RNA polymerases.

87. TBP is the only **DNA binding protein** that binds in the **minor groove,** while other **protein binds DNA in the major groove.**

88. The ability of TBP to bind DNA is controlled by one of the TAFs.

89. TATA Box has the **core sequence** of TATAA.

90. TATA Box is usually surrounded by **G-C rich sequences.**

91. **CAAT Box** is located about 80 bp upstream of the start point.

92. **CAAT Box** play a strong role in **promoter efficiency.**

93. **Promoters for polymerases III.**

94. The downstream or internal promoters occur in **5 sRNA & tRNA genes.**

95. Promoters and enhance share the modular organization & some of the elements present in them are common, therefore an enhances may be regarded as a part of the **promoter located at a distant position from the start point.**

96. A complex of proteins formed at the promoter as a prelude to initiation of transcription is called **Transcription complex.**

97. In the virus **SV 40,** termination site consists of a **hairpin loop, followed by a stretch of U bases.**

98. The termination site is a run of 4 U bases with in a **G-C rich region,** but a terminator with 5 U bases is more efficient.

99. The critical feature of termination is the recognition of a sequence of 4 U bases within a **G-C rich region.**

100. The largest sub-unit of RNA polymerase is degraded by the **proteasome.**

101. Activation of transcription is associated with acetylation of **histones,** particularly of H3 & H4, in their N- terminal tails while transcription is associated with **deacetylation.**

Processing of rRNAs:

102. The 1st processing reactions seems to be cleavage of the **superfluous** leader sequence at the **5' end.** The next cleavage occurs within the **spacer sequence** located between the 18 S & 5.8 S sequences.
103. An abundant component of nucleolus- **Fibrillarin protein.**
104. snoRNA U3 is required for the 1st cleavage events in the 5' transcribed spacer region in both **yeast & xenopus.**

Processing of tRNAs:

105. The 3' terminus of tRNA molecule is generated by **RNase D,** which degrades the 3' end till.
106. **Exons** are the sequences that form the mature mRNAs.
107. **Introns** are deleted during processing of pre mRNAs.
108. The excision of intron and the joining together of all the exons of a gene in a proper sequence to yield the **mature mRNA** called **splicing.**
109. The primary RNA transcribed of a structural gene is called **pre-mRNA** from which mRNA is generated by splicing.
110. Introns present in genes except **tRNA exceeding nuclear genes.**
111. **Transesterification** reaction is defined as a reaction in which one phosphate ester is converted into another phosphate ester without any intermediary hydrolysis.

Translation:

112. Conversion of mRNA base sequence into protein amino acid is called **Translation.**
113. **A bacteria cell** contains about 20,000 ribosome, about 1500 mRNA and about 2×10^6 protein molecules.
114. The number and the sequence of bases in mRNA specifying an amino acids is known as **codon.**
115. The set of bases in a tRNA that base pairs with the codon of an mRNA is called **anticodon.**
116. The set of all the codons that specify the **20 amino acids** is termed as the **genetic code, genetic language or coding dictionary.**
117. A large (44 codon) of codons will not code for any amino acids, called non-sense codon, else more than one codon will code for a single amino acid. Such a code being known as **redundant or degenerate.**
118. Non-sense codon- UAA, UAG, UGA; these codon acts as polypeptides terminations codons.
119. The codon AUG acts as the initiator codon; in prokaryotes its codes for **formyl methionine** at the beginning of a polypeptide.
120. Genetic analysis of detection and insertion mutants of the **Beistron** of the rII of phase T_4 was reported by **Crick *et al.* (1961).**

121. The 1st synthetic RNA used for protein synthesis in a cell free system was **poly U;** it directed the synthesis of **poly phenylacanine (UUU).**

122. A triribonucleotide cannot direct the synthesis of a polypeptide but it leads to the binding of specific **amino acyle-tRNA** to the ribosomes. This property was exploited by a **Nirenberg-leder technique.** For assigning specific codons to specific amino acids.

123. Tri-nucleotides with 5' terminal phosphate are active in inducing the binding of **amino acyle tRNA** to ribosomes; while diribonucleotides are inactive.

124. Crick (1966) proposed **Wobble hypothesis** to account for the degeneracy of genetic code. Acoording to this theory "I[st] two base pair (codon) pair according normal base pairing role with the last two bases of the anticodon."

125. **Khorana technique** to confirm that the code is **commaless overlapping & degenerate and each codon is a triplet.**

126. The Wobble hypothesis can explain **degeneracy** only at the **third position of the codon** & at best this can account **only for three codons for each amino acids.**

127. The ribosome control the environment in such a way that **A pairs with U & G pairs with C** (conventional pairing) at the I[st] two position of the codon.

128. At internal sites of **mRNA,** AUG specifies methionine, GUG is read as **Valine,** while UUG means **leucine.**

129. In **eukaryotes, mature mRNAs** are translocated to the **cytoplasm where translocation takes place.**

130. At 37°C, transcription in bacteria proceeds at **~40 nucleotides/sec. (~14 codon/second), while the translation rate is ~15 amino acids/seconds.**

131. **Functional half life** is based on the **ability to function as mRNA & support translation.**

132. **In eukaryotes,** mRNA survives 40 to 60 times as long as prokaryotic mRNA.

133. **The mRNA present in seeds survive for years *e.g.* Lotus** seeds are known to germinate after **1,700 years.** Since, there is no RNA synthesis during germination, but there is active protein synthesis, the mRNA must have survive for 1700 years in these; seeds such mRNA are known as **Long-lived mRNA.**

134. **A single** prokaryotic mRNA molecule codes for more than one polypeptide, such **mRNA is known as polycistronic mRNA. In facts, all genes located on a single operon.**

135. Intercistronic sequences are obviously not **translated.**

136. Functional mRNA **dissociate as ~ 18 S rib nucleoprotein (RNP) particles following EDTA treatment of eukaryotes poyribosomes.**

137. All eukaryotic mRNAs have a 5' cap but cap is absent from mRNA in mitochondria and chloroplasts.

138. The 5' leader sequences is usually short (< 100 bases) but the 3' trailer is long (upto 1000 bases).

139. Poly A (tails is associated with 70 KD protein, the **poly (A) binding protein,** which is the predominant protein associated with **active mRNA.**

140. Bacterial mRNAs have a common **consensus hexamer sequence, called shine-Dalgarno sequence (AGGAGG), ~7 bases upstream of the AUG codon.**

141. The UCCUCC consensus of 16S rRNA base pairs with AGGAGG consensus of mRNA; this allows the similar sub-unit of ribosomes to bind to mRNA.

142. **In eukaryotes,** translation initiation is based on **recognition on the** 5'-cap of mRNA by some **cap binding proteins.**

143. When the 40S sub-unit recognize an **AUG, the 60 S sub-units of ribosomes attaches to it & translation begins.** This is referred to as the **scanning model of translation initiation.**

144. The 40 S sub-unit associated with upstream site & migrates to the downstream site having **AUG codon.**

145. **Non-sense codons or termination codons-** UAA, UAG, UGA which do not code for any amino acids.

146. Bacterial mRNA is **degraded by a combination of endonuclease & exonuclease,** degradation progressing from **5' to the 3' end of mRNA.**

147. Endonucleases make a succession of internal **cleavages** in an **mRNA** molecule following the last ribosome translating this mRNA to results generates a series of fragments which are degraded by **exonucleases from their 3' OH end.**

148. Degradation of mRNA may be catalysed by a complex often called **degradosome, having RNase E, PNPase (polynucleotide phosphorylase) & a helicase.**

149. The helicase unwinds the substrates mRNA to make it available to PNPase; which is 3'→ 5' exonuclease.

150. **Polyadenylation of mRNA promotes degradation.**

151. **Exosomes is also involved in the processing of rRNA precursor.**

152. tRNA is a class of RNA of small size of 3S & different tRNAs having 74 to 95 nucleotides; most commonly, tRNA has 76 nucleotides.

153. **The distance of anticodon from the 3' amino acid acceptor end is uniformly 66 Å.**

154. The clover leaf model represents the **secondary structure of tRNAs.**

155. The clover leaf is stabilized by base-pairing involving H bond called **secondary H bonds.**

156. The base paired double helical stems of the secondary structure are maintained in the tertiary structure.

157. All the different tRNAs which are normally charged within the same amino acid are called **iso-accepting tRNAs.**

158. **Reading frame** refers to the I[st] base of the sequence from which reading begins and proceeds 3 bases at a time. Therefore, each DNA strand has three possible reading frames; & a duplex has six possible reading frames.

159. Albinism is recessive condition caused by a defect in **phenylalanine trysine metabolism.**

Gene regulation in Prokaryotes:

160. **The Operon model** was Ist described in **1961** by **F. Jacob & J. Monod** and also received Noble Prize in 1965.

161. **Operon model to explain** the regulation of genes encoding the enzymes required for **Lactose utilization in *E. coli* &** proposed that the transcription of one or a set of contiguous structural genes (gene coding for polypeptide) is regulated by **two controlling elements, called regulator genes or repressor gene,** codes for a protein called repressor bind to the second element called operator genes.

162. **Operator** is usually located between the **promoter and structural genes.**

163. The Lac operon contains **a promoter, an operator and three structural genes** Z, Y, α coding for the enzyme **β galactosidase, β galactoside permease and β galactoside transacetylase, respectively.**

164. The Lac regulator gene, designated the **i-gene, codes for a repressor *i.e.* 360 amino acid length**

Molecular Markers:

165. Marker is a DNA sequence which is readily selected & whose inheritance easily be monitored.

166. The use of molecular marker is based on naturally occurring DNA polymorphism.

167. Genetic polymorphism is defined as the simultaneous occurrence of the trait in the same population of two or more discontinuous variants or genotypes.

168. The Ist DNA markers to be utilised were fragments produced by restriction – digestion (RFLP – Restriction Fragment Length Polymorphism).

169. **Co-dominant inheritance:** The different forms of marker, should be detectable in diploid organism to allow discrimination of homo & heterozygotes.

170. Non PCR based approaches: **RFLP**

171. **PCR based approaches:** RAPD, SSRP, AFLP, AP-PCR, SCAR, SNP etc.

172. Targeted PCR & sequencing based: **STS, SCARs, STMs, CAPs.**

173. The Ist DNA based genetic markers were the Restriction. Fragment Length Polymorphism markers used by **Botstein *et al.* (1980).**

RFLP:

174. RFLP was the Ist technology which enabled the detection of polymorphism at the DNA sequence level.

175. The DNA sequence variation detection by this method was termed restriction fragment length polymorphism.

176. Genomic restriction fragments of different lengths between genotypes can be detected on **southern blots & by a suitable probe.**

177. Genomic DNA of an organism is isolated and digested with a particular **restriction exonuclease.**

178. Recombination frequencies can also be estimation from doubled haploids (DH) derived from the **pollen of F_1 plants.**
179. Doubled hybrid populations consist of completely homozygous, true breeding individuals.
180. In maize, recombinant inbred lines (RILs) have also been produced from a number of crosses through **continuous selfing or sib-mating.**
181. Very polymorphic plant such as inbred lines of maize usually require **one or two** enzymes to detect polymorphism with most probe but in case of Rice, 11 enzymes were used & about 75% of the probes were polymorphic with these.
182. RFLPs are **co-dominant markers**, enabling heterozygotes to be distinguish from homozygotes.
183. RFLP is simple as **no sequence specific information is required.**
184. RFLPs, used in direct selection using qualitative traits.
185. **Linkage between RFLP and QTL can also be used for the prediction of combining ability (GCA).**
186. RADP & RFLP markers are used for the indirect selection of segregating population.
187. RADPs have been utilised for the identification of **somatic hybrids.**
188. RADP is sensitive to changes in PCR conditions, resulting in changes to some of the amplified fragments.
189. The SAMPL (Selective Amplification of Micro-satellite polymorphic loci) primer is based on the sequence of two different adjacent SSRs.

DNA finger printing:

190. DNA finger printing methods have been widely used in **criminal investigation** & for the determining the pedigree & parentage of an individual.
191. The technique of classical DNA finger printing is a **derivative of RFLP analysis.**
192. AFLP approach is widely used for **developing polymorphic markers** used in detection of polymorphism between closely related genotypes.
193. A hybridization based approach is **oligonucleotide finger printing.**
194. Detection of hybridization/hybridizing fragments (fingerprints) by **auto-radiogrpahy or by chemiluminesence.**
195. Micro-Satellite complimentary oligonucleotides are also useful for banding pattern when hybridized to PCR products generated by **arbitrary primers.**
196. A modification of RAPD **is DAF (DNA Amplification Fragments)** has been shown to generate reliable finger prints because many more bands are resolved due to **short primers.**
197. Nei & Li coefficient (N) has been recommended for analysis of **multilocus finger printing.**
198. **DNA finger prints** are also be used for the **genetic mapping.**
199. Multilocus DNA profiles are usually translated into **discrete characters** ***e.g.*** "0" for absence & "1" for presence of band at a particular position.

200. Bands are generally inherited in a Mendelian fashion an application of **RFLP finger printing** is the relatedness between parents & off-springs.

MAS (Marker Assisted Selection):

201. Markers should co-aggregate or to be closely linked with the desired traits.
202. **NILs (Nearly Isogenic Lines):** The BC_7 F_2 line is said to be nearly isogenic lines with the recipient parent.
203. **Bulk Segregant Analysis (BSA):** It requires the generation of populations of bulk segregants (Bulks).
204. Favourable QTL alleles can be used to transfer one or more discrete genomic segments from a donor to an elite cultivar by **back-crossing.**
205. **MABC (Marker Assisted Back Crossing),** in which a chosen allele at a marker locus is transferred from a donor to a recipient line, has been widely used to introgress favourable alleles into elite material which lacks a specific trait.
206. A approach aims direct recombination between selected individual as part of a breeding scheme seeking to generate an ideal genotype. This variety development approach is commonly referred to as **marker assisted recurrent selection (MARS) or genetic construction.**
207. RFLP markers have been linked to **Linolenic acid** content fan locus in soyabean.
208. RADP markers that control **somatic embryogenesis in alflalfa.**
209. Photoperiod sensitive genic male sterility genes in rice designated PMS1 & PMS2 were located on chromosome number 7 & 2 respectively (Zhang *et el.* 1994).
210. The transferred DNA may be expressed for only a short period of time following the DNA transfer process & it's called **transient expression.**
211. Transient assays are commonly used for the analysis of gene expression and rapid monitoring of gene transfer.
212. Marker gene is catagorized into two parts:
    (*a*) Reporter or screenable genes.
    (*b*) Selectable marker genes.
213. Ideal reporter genes have few features:-
    (*a*) Detection with high sensitivity
    (*b*) Low endogenous background
    (*c*) There should be quantitative assay.
214. **Octopine synthase or nopaline synthase** (reporter genes) are present in the T-DNA of T: (tumor inducius) or Ri (Root inducing) plasmid of Agrobacterium.
215. **Chloramphenicol Acetyl Transferase** is one of the most commonly used reporter genes in **eukaryotic organism. (Found in Tn 9).**
216. The presence of acetyl chloramphenicol is detected through **auto radiography.**

217. **Chloramphenicol acetyl transferase was the I[st] genes** to be introduced into plant cell (Herrena-Estrella *et al.* 1983).

218. Bacterial luciferase (Lux 2) have originated from ***Vibrio harveyi & Vibrio fischerii.***

219. The Vibrio harveyi Luciferase, is encoded by the **Lux A & Lux B cistrons** that are part of a longer operons.

220. Light is measured as **photons or quanta/second.**

221. Anthocyanin's are prominent red & purple pigments of flowering plants.

222. Anthocyanin have been used as visible marker for transformation.

223. The rol gene derived from ***A. rhizogenes*** is responsible for the proliferation of hairy roots.

224. Agrobacterium are gram negative rods belong to the bacterial family of **Rhizobiaceae.**

225. **Zaenen *et al.* (1974) I[st] noted** that virulent strain of Agrobacterium tumefaciens.

226. The large plasmid in agrobacteria is called **Ti (tumor inducing bacteria) & Ri (Root inducing bacteria), respectively.**

227. **Ti** plasmids are defined as octopine, napoline, succinamopine, agropine or leucinopine.

228. Ti plasmids of agrobacterium are large circular, DNA molecules, up to 200 kb in length with molecular weights of about 1.2×10^8 of agrobacterium chromosome.

229. The T-DNA (transferred DNA) encodes for both **auxin & cytokinin** biosynthesis in the plant genome, which alerts with the hormone balance in the plant cell & lead to formation of **gall phenotype.**

230. An **opine synthesis of** os region is responsible for the synthesis of unusual amino acid or sugar derivative, which are collectively given the name **opines.**

231. Opine synthesis is controlled by **enzyme octopine synthase and nopaline synthase at loci called** ocs and nos on octopine & nopaline plasmids, respectively.

232. Opine catabolism is controlled by **Loci noc & occ for** nopaline & octopine catabolism, respectively.

Recombinant DNA Technology:

233. In 1995, the variety of tomato named as **Flavr Savr** was given regulatory approval for commercial cultivation which had delayed ripening gene.

234. The four principal transgenic crops in the world are **soyabean, maize, cotton & rapeseed or canola.**

235. The **USA is one** of the six "founder biotech crop countries" having commercialized biotech **maize, soyabean, cotton & potato in 1996.**

236. NCFAP – National Centre for Food & Agriculture Policy.

237. **In 1988, Novo Nordisk** was the 1st company to commercially produce a fat **degrading enzyme.**

238. Breakage of a DNA segment (molecule) at the desired places to isolate a specific DNA segment & then insert it in another DNA molecule at desired position (product called **recombinant DNA**).

239. The restriction enzymes were discovered for the 1st time by **Arber (1962), but purified later in 1970 by D. Nathans & H. Smith.** In 1978 received an Noble Prize by D. Nathans & H. Smith on his pioneer work.

240. Different cleavage enzyme recognize different but specific sequences, each ranging in length from **4 to 8 bps.**

241. Based on attributes, restriction enzymes have been grouped into three classes:

(*a*) **Type II restriction enzyme (EcoRI)** use in cleaning

(*b*) **Type I (EcoK & EcoB)**

(*c*) **Type III (EcoPI & Hinf III)**

242. The technique of **gel electrophoresis** is used for separation of DNA molecules of different sizes.

243. A new technique called **Pulsed Field Gel Electrophoresis (PFGE)** has been used for separation of large size DNA molecules, sometimes representing whole chromosomes.

244. A mixture of DNA, RNA or protein fragments can be separated be **gel electrophoresis** & the separated bands can be stained & visualized directly in the **gel.**

245. The DNA bands in the gel are denatured into single strand with the help of an **alkali-solution.**

246. The technique of southern blotting used for DNA transferred from "gel" to the membrane could not be used for blot transfer of RNA.

247. The mRNA bands blotted onto **nitrocellulose membrane can be hybridized with a labelled DNA or RNA probe.**

248. The single stranded regions of the probe are removed by **nuclease (Mungbean nuclease or S-1 nuclease).**

249. **Western blotting technique** used for detection of protein.

250. **pBR 322** is a vector contains resistant against two **antibiotics** due to **two genes, one for resistant against** arnpicillin and **other resistance** tetracycline.

251. pBR 322 is derived from *E.coli* plasmid **Col E1.**

252. EMBL 3 & EMBL 4 are called **replacement vectors.**

253. **MI 3** as a vector for **DNA sequencing;** contain a 7.2 kb long single stranded circular DNA. MI 3 is a filamentous bacteriophase of ***E. coli.***

254. Transposons of ***Drosophila*** **(P elements)** are alo used as a vectors.

255. **Yeast Artificial Chromosomes (YACs)** as a vector have been constructed utilizing the yeast centromere sequences, telomere sequences, autonomous replicating sequences & other necessary DNA sequences.

256. **BACs (Bacterial Artificial Chromosomes) and MACs (MammalianArtificial Chromosomes);** are useful for a variety purposes but are to big & are difficult to use because they need to be multiplied in yeast & mammalian cells, respectively.

257. **Chimeric DNA molecules** obtained by the inserting of a foreign DNA segments into the DNA molecules of the vectors.

258. **Chimeric DNA molecules mean a small DNA sequences.**

259. **Bacteria** can be classified into

    (*a*) Gram negative *e.g.* ***E. coli, Pseudomonas & Rhizobium* etc.**

    (*b*) Gram positive *e.g.* ***Bacillus subtilis & Streptomyces spp.***

260. Chromosomes found in the nucleus in eukaryotes are separated from the rest of cell through nuclear membrane.

261. **Simian virus 40 (SV 40)** is one such virus, in which **globin** gene could be integrated. This gene integrated in SV 40 could be transcribed & translated in **mouse kidney cells.**

262. Adeno virus, retroviruses are also used for **gene therapy in human being.**

263 **cDNA (complementary DNA)** libraries can be prepared by **isolating mRNA from tissues** which are actively synthesizing proteins, like root & leaves in plant.

264. The mRNA used for copying it into cDNA by use of **reverse transcriptase.**

265. cDNA clones will differs from genomic clones in lacking the **introns** present in **split genes.**

266. **The technique of recombinant DNA & gene cloning** permit us to obtain an unlimited supply of identical copies of a gene sequence or DNA segments.

267. In 1985, developed PCR technique.

268. **Nick name of PCR-** Peoples choice reaction.

269. Temperature of denaturation of DNA – **90-98°C**

Major differences of PCR & Gene cloning

S.No.	Parameters	PCR	Gene cloning
1.	Manipulation	Invitro	in vitro & in vivo
2.	Selectivity of specific segment from complex DNA 1st step	Last step	
3.	Biological reagents required	**DNA, Taq polymerase, d NTPs**	**Restriction enzyme Ligase, vector, bacteria cells**
4.	Automation	Yes	No

270. **Taq-** Thermus aquaticus

271. The enzymatic progress best temperature **72ºC & denaturation 90ºC.**

272. **PCR is also extensively used for the study of DNA polymorphism with the help of a variety of molecular markers.**

273. PCR based markers – RAPDs, SSR, SNP, AFLP

274. Molecular maps (both genetic & physical) are now available for all major crops like **maize, tomato, potato, rice, barley, wheat etc. and** also for animals like **Drosophila, mouse and humans.**

275. The genetic maps prepared through estimation of recombination frequencies do not extractly correspond to the physical feature either on the chromosome or on the DNA molecules.

276. The fine structure of gene can be studied only through determination of nucleotide sequence of the concern DNA segments.

277. **The restriction map can be used for determining the nucleotide sequence of a specific gene.**

278. Restriction maps of BACs (Bacterial Artificial Chromosomes) have also been used for preparing **BACs based physical map of chromosome.**

279. Molecular markers have been used for the preparation of molecular genetic maps. The markers are 1st screened for detecting polymorphism between the parents, that are used for the production of the mapping population **(like F_2, BC_1, DH, RILs).**

280. The specific cloned sequences of DNA **can be used to identify variation at the DNA levels.** These variations are monitored as changes in the length of defined individual DNA fragments produced by digestion of DNA samples with **restriction endonuclease. These variation called RFLPs.**

281. **RFLPs variations (markers) have been used for genetic mapping** in human beings, mice, fruitfly & crops.

282. **RAPDs markers** demonstrating genetic diversity in maize genotypes.

283. RAPDs technique may also be combined with **RFLPs to increase the efficiency of molecular marker for genetic analysis.**

284. RAPD technique has been found to be inefficient due to its **poor reproducibility.**

285. **Simple sequence repeats (SSRs)** represent variation in the repeat number of sequences, 1-6 bps long. Micro-satellite maps have been prepared in **mouse, human, barley, wheat** *etc.*

286. Molecular maps are also used for **map based cloning** of a specific gene.

287. **Paul Berg (Noble Prize)** in 1991 redefined **reverse genetics.**

288. The cDNA clones may also be used **directly for gene manipulation and transformation experiments.**

289. Genes for storage proteins are expressed only in developing seed, **ovalbumin gene is expressed in oviduct & globin gene expressed in erythrocytes.**

290. The gene for **chalcone synthase (CHS)** has been isolated from ***Antirrhinum majus*** and **Petunia hybrid** using **heterologous cDNA (cloned DNA or complementary DNA or probe).**

291. **Heterologous probe (Antirrhinum cDNA)** was used for isolation of **chalcone (CHS) synthase** gene from barley.

292. In maize, probe used for isolation of barley genes **Wx (waxy gene) & Al (aleurone gene).**

293. A number of genes with unknown genes products are expressed in specialised tissues. Isolation of such gene can be achieved by **differential screening.**

294. One of the limitation of differential screening, however, is that the technique can be used only those genes that have high expression & may not prove suitable for those differentially expressed genes that have low expression (differential display).

295. **Frederick Sanger (who won noble prize twice)** had initially developed a method for gene/ DNA sequencing, which utilized DNA polymerase to extend **DNA chain length**. This was termed **plus-minus method.**

296. **DNA polymerase I** is called **sequenase** is used due to its high processivity in sequencing of DNA.

297. **A precursor molecule** is 1st synthesized which subsequently after losing segment of RNA by **cleavage,** give rise to **tRNA.**

298. Interferon is a protein, which has a property of inhibiting viral infection & cell proliferation. It can be used for treatment of **viral infection & malignancies (cancer).**

299. In 1980, **Weismann & co-workers published the nucleotide sequence the 1st** IFN-α has not been intron sequencing.

300. **In 1970, H. Temin & D. Baltimore** discovered the presence of RNA directed DNA polymerase enzyme which has **ability for synthesizing DNA or RNA template.**

301. Gene transfer technique, utilising recombinant DNA technology differ for animals & plants. While in animals, **DNA is injected into either the pronuclei of embryonic cell (fertilized egg) or the embryo stem cells in plants, cells, protoplasts or meristemic embryonic tissues are used.**

302. Transfection or gene transfer in animals can be carried out at the cellular level. **(molecular farming).**

303. The transfection of fertilized egg may involve transfer of **whole nuclei, whole chromosomes or DNA segments.**

304. For the transfer of whole nuclei the egg cell are treated with **cytochalasin-B and subjected to centrifugation causing enucleation** (nucleus is removed).

305. During 1997, reported that the utilized for **transfer of a somatic nucleus to an egg** leading to the production of a clone or sheep named **Dolly.**

306. **In case of Ti & Ri plasmids,** 'vir' region having six operons **vir A, B, C, D, E & G,** to responsible for virulence towards host & **T DNA region.**

307. The most commonly used selectable marker is the gene encoding **neomycin phosphotransferase (npt II)** importing resistance to **Kanamycin.**

308. **For gene transfer,** the explants commonly used include protoplasts, suspension cultured cells, callus cells clumps, thin cell layers, tissues slices, leaf discs and section of roots, stem or floral tissues.

309. **Agrobacterium mediated,** gene transfer was extensively utilized for production of transgenic **dicotyledonous plants;** initially it was not successful in **monocotyledonous plants.**

310. Agrobacterium mediated transformation in monocots including cereals like **maize, rice and wheat.**

311. **Microprojectiles** technique has been successfully used in **soyabean, tobacco, maize, rice, wheat, oat.**

312. The 1st transgenic plants were produced in **Europe & USA in 1983.**

313. Through transgenic approach, improvement in **vitamin A/C/E** (most important of these is the so called **Golden rice).**

314. Development of **cafferine free transgenic coffee.**

315. Gene for nucleocapsid protein (N) was transferred from some **tospovirus like tomato spotted wilt virus** & provided resistance against tospoviruses in **tomato, tobacco, groundnut, pepper etc.**

316. In tomato using Antisense RNA technology, transgenic plants were produced, which were either **bruise resistant (suitable for transport & storage) or exhibited delayed ripening (flavour Sour tomatoes).**

317. During 1990-92, genes **barnase (encoding a ribonuclease) & barstar (encoding a ribonuclease inhibitor),** which derived from **Bacillus amyloliquefaciens,** were separately transferred to ***Brassica napus*** to produce male sterile & male fertile plants, respectively.

318. **The gene constructs were prepared, where** barnase/barstar was fused with **TA29 promoter** from tobacco, permitting expression only in the anther, causing male sterile and male fertile restoration, respectively.

319. **Golden mustard,** rich in vitamin A.

320. Transgenic tobacco plants with increased level of **mannitol** were produced using a gene for **mannitol dehydrogenase** (this also gave resistance against **salinity**).

321. Monoclonal antibodies can be produced in a specialised cells through a technique called **hybridoma technology,** was discovered by **Georges Kohler** of West Germany and **Cesal Milstein of Argentina in 1975;** who jointly with **Neils Jerne** of Denmark were awarded **Noble Prize for medicine of physiology (1984).**

322. **In human,** α-like protein (globin) genes are located on **chromosome 16** and β-like globin genes are located on **chromosome 11.**

323. **Feto protein** is a major component of a blood plasma.

324. Chorion protein, a major component of egg cell in insects.

325. rRNA genes are present in multiple copies at **Nucleolar Organizing Regions (NORs)** of specific satellite chromosomes.

326. In human small nuclear RNA or SnRNA **U1 there are 30 functional genes located on chromosome 1 & 500 to 1000 pseudogenes distributed throughout the genome.**

327. Discovery of **Olfactory Receptor (OR) Genes** led to the award of 2004 Noble Prize in Medicine or physiology to **Richard Axel and Linda B. Buck (both of USA).**

328. Haemoglobin is a multimeric protein consisting of **four sub-units, two α sub-units & two β sub-units.**

329. α sub-unit found on chromosome 16 and β sub-unit found on chromosome 11.

330. The endoplasmic reticulum regions carrying ribosomes are called **rough endoplasmic reticulum** & those devoid of ribosome called **smooth endoplasmic reticulum.**

331. Smooth endoplasmic reticulum region is used for the formation of **transport vesicles,** which carry protein & lipids to the golgi bodies.

332. Smooth endoplasmic reticulum region rich in glycogen.

333. **Cytosolic surface is used for synthesis of majority of proteins.**

334. Methylation of normal bases is the most common method giving rise to these (pseudo uridine) unusual bases.

335. All tRNA molecules had **guanine residue "G" at** 5' end & unpaired CCA sequence at 3' end.

336. **Three dimensional structure** of tRNA could be studied through **X rays crystallography** as was done for DNA double helix. A. Klug from MRC laboratory, Cambridge, who was awarded Noble Prize in **1982** in chemistry to study of TDS at tRNA.

337. In 1973, S. H. Kim suggested that T. D. S. of tRNA tale the shape of letter **L with thickness of 20Å, & derived from two dimensional clover leaf model.**

338. **Pre-initiation** of translation includes formation of amino acyl tRNAs & dissociation of ribosome into its two sub-units.

339. Formation of amino acyl tRNA consist two steps:-

(*a*) Activation of amino acid (*b*) transfer of amino acid to tRNA.

340. Ribosomes, engaged in protein synthesis & associated with **mRNA as 70S (or 80S) particles, are dissociated into two sub-units (30S, 50S)** with the help of the protein called **initiation factor 3** in prokaryotes and **eIF 6** in eukaryotes.

341. Translation initiation involves interaction of mRNA & tRNA with **30S** sub-unit of ribosome, **formylation of met-tRNA met and association of the two sub-units of ribosome leading to the formation of** 70S initiation complex. Three initiation factors- IF 1, 2, 3 involves in its proceduce.

342. In prokaryotes a termination of codon is recognized by one of the two release factor, **RF1 & RF2.**

343. RF1 recognizes UAA, UAG.

344. RF2 recognize **UAA, UGA.**

345. **UAA recognized by both RF1 & RF2.**

346. **In 1975,** protein synthesis is isolated chloroplast of **pea was** studied by **R. J. Ellis in U. K.**

347. Proteins synthesized in mitochondria have also been identified by **inhibiting protein synthesis in cytoplasm** by cycloheximide.

348. Adenosine triphasphatase (ATPase) is an important component of mitochondrial membranes, which plays an important role in coupling respiration to ATP formation.

349. Position effects represent a change in the function of a gene due to its location in a chromosome or in the homologous chromosome.

350. Variegated position effect involve **heterochromatinization, which prevent the transcription of concerned genes.**

351. **Example** of variegated position effect is provided by the **mottled eye mutant in *Drosophilla*.**

352. Presence of an additional Y chromosome in *Drosophila* reduces the frequency of **heterochromatinization.**

353. The formation of heterochromatin occurs in two steps:

(*a*) **Nucleation:** it occurs at a specific DNA sequence.

(*b*) **Propagation:** spread of heterochromatin organization.

354. The phenotypic effect of **apr and w genes in heterozygotes** state depends on their **cis-trans position,** giving this phenomenon the name **cis-transposition effect.**

355. **Position effect produce by duplication,** resulting from unequal crossing-over is clearly demonstrated in the case of **Bar (B) eye gene of Drosophila.**

356. In Ac-Ds system, Ds (responding element) may be located near the **structural gene C_1 on chromosome 9 & C_1 produced color giving rise to colored kernels.**

357. Ds in maize suppressed the color of kernels in maize.

358. The responding element **Ds appears to be produced by an internal deletion in the Ac.**

359. Paramutation refers to a change in the function of an allele (paramutable allele) of a structural gene that may last for several generations and its induced by the presence in the same nucleus with another allele.

360. Exo **gene R** is in chromosome 10 of maize produces full **anthocyanin in aleurone & certain vegetative parts.**

361. **Paramutated gene state of an allele is induced by a** paramutagenic allele, while the recovery in the paramutated allele is **spontaneous.**

362. An **epigenetic change** is an alteration in gene expression associated with alternative chromatin or methylation states without any change in **DNA sequences.**

363. Paramutagenic allele cause an increase in the number of copies of a heterochromatic sequence, **called metamere;** associate with mutable allele.

364. The sequence of paramutable allele is change by the gene conversion.

365. **A maize gene (MOP1, mediator of paramutation 1) has been identified that is required for establishment and maintenance of the paramutagenic state of several loci.**

366. Isopropyl thiogalactoside (IPTG) is not recognized by β galactosidase, such inducers, which induce enzyme synthesis without getting metabolized are called **gratuitous inducers, which are very useful for conducting recombinant DNA studies.**

367. F. Jacob & J. Monod in 1961, on the basis of their study on inducible system for the synthesis of β galactosidase enzyme in *E. coli.* Proposed a model to explain the induction or repression of enzyme synthesis **(operon model).**

368. Operon model assumed the presence of a unit of co-ordinated control of protein synthesis & this unit was called **operon.**

369. **Operon consisted-** operator gene (which controlled activity), structural gene (which take part in the synthesis of protein). This means that the structural genes will synthesize **mRNA under the operational control of an operator gene.**

370. Example of two different operons – **Lac & tryptophan in bacteria.**

371. Jacob & Monod themselves, who shared the **Noble Prize with** A. Lwaff **for their work on "genetic control of enzyme & virus synthesis".** The experiment is popularly described as **Pa Ja Mo experiment.**

372. The Z gene was expressed (through the synthesis of β galactosidase) very fast and at maximum rate from the beginning, leading to the formation of **mRNA hypothesis.**

373. **Lac operon a group of genes found in *E. coli.***

374. Lac operon consists of 26 base pairs and is a part of the 1st structural gene **Lac Z (meant for β galactosidase).**

375. Lac operon in *E. coli*, consists three genes (structural gene).

(*a*) Gene Lac Z (3063 bp) **codes for the** β galactosidase which **active as a tetramer & breaks lactose into glucose and galactose to be utilized in the cell.**

(*b*) Gene Lac Y (800 bp) **codes for** β galactosidase permease, **which is a membrane bond protein & help in transport of metabolites.**

(*c*) Gene Lac A (800 bp) **which codes for β galactosidase transacetylase, enzyme that transfers an acetyl group from acetyl-Co A to β galactosidase.**

Note:- **DNA double helix is negative charge of e^- particles.**

376. In Lac operon of *E. coli*, repressor is synthesized through activity of a gene **Lac I, known as regulator gene.**

377. Repressor then attaches on operator gene & checks synthesis of mRNA.

378. A promoter gene located upstream **(5' end)** controls the activity of the regulator gene.

379. The bound protein in inactive repressor is **allolactose.** It has also been shown that while β galactosidase breaks lactose into glucose and galactose, a side reaction changes galactose into allolactose & galactobiose.

Promoter gene:-

380. Operator gene is the site of action of repressor, the actual site transcription start is different & is known as **promoter gene; which lies upstream of the operator gene.**

381. The repressor, when attached to operator, prevents the movement of **RNA polymerase.**

382. **Catabolite activator protein** associated with the Lac operon, there is another site, 16 bp upstream of the promoter, **which is used for a positive control gene expression.**

383. Two proteins for the regulation of **Lactose operon:-**

(*a*) Lac repressor, (*b*) Catabolite gene activator (CAP)

384. Repressor binds to operator site & **CAP or CAG** protein binds to cga site.

385. A protein such as CAP, which interacts with **DNA & RNA polymerase** to assist in transcription initiation, is called **Apo inducer.**

386. Cyclic AMP is essential for activation of cga protein, since when cyclic AMP was removed or absent due to **mutation. Transcription was reduced.**

387. Level of glucose concentration influences this **cyclic AMP cga protein system.**

388. Transaction in lac operon requires that

(*a*) CAP- positive control

(*b*) Lac repressor- negative control

389. The binding of purified lac repressor protein to DNA was 1st characterized by a **filter binding assay.**

390. **Lac repressor** is a polypeptide of **360 amino acids &** is divided into two regions.

(*a*) 1-59 amino acids reduce on the amino terminal end make **a long head piece** which can be cleaved at position 51, so that the fragment 1-51 is called **short head piece.**

(*b*) 60-360 amino acids reduces on the C-terminal end make **trypsin resistant core** (trypsin cleaves the repressor at amino acid no. 59).

391. **CAM protein** is a homodimer, consisting of two identical polypeptides, each consisting of **209 amino acids** reduces with two domains. The carboxyl terminal domain of polypeptide helps in binding to the cga site & also to the α sub-unit of RNA polymerase enzyme **(α-CTD domain).**

392. The larger amino acid terminal domain is **responsible for dimerization as well as for c-AMP binding.**

<u>Tryptophan operon in bacteria (*E. coli* & salmonella):</u>

393. Five structural genes that are involved in the synthesis of tryptophan form an operon with an operator, & code for three enzymes needed for tryptophan synthesis.

394. The major differences between lactose and tryptophan operon is that tryptophan increases the affinity of repressor protein for the operator, while **leucine** (the inducer for lactose operon) inactivates the repressor.

395. The lac repressor acts only on lac operon with **Z, Y, α** cluster of genes.

396. Trp repressor control following three sets of unlinked genes:

***(i)* Tap EDBCA** meant for tryptophan synthesis enzymes

***(ii)* aro H gene** coding for an enzyme needed for common pathway of aromatic amino acids bio-synthesis.

***(iii)* Trp R regulator gene**

397. **Leucine regulatory protein (Lrp),** which responds to the presence of leucine & regulates the expression of many operon which are collectively described as **Leucine/Lrp regulon.**

398. Lrp is described as **global regulator of gene expression in *E. coli* & other several bacteria.**

399. Lrp is a DNA binding protein, **which does not bind single stranded DNA.**

400. Lrp is a transcriptional regulator, when bond with leucine it functions as an activator & when free it function as repressor.

401. The **arabinose operons (also called as OBAD operon) have** positive or negative controlled system found in *E.coli.*

402. Upstream to the starting point of Lac operons are number of sequences which are common to many *E.coli* and bacteriophage genes.

403. There is a sequence at about 10 bases upstream from starting point (also called-10 sequence) called **Pribnow Box** which is an important sequence & named after **D. Pribnow,** who first discovered it in **1975.**

404. Pribnow Box–(TATAAT).

405. In tryptophan operon of *E.coli*, the mRNA of gene trp E (gene for enzyme **anthranilate synthetase**) has a sequence of 162 bases between its initiation codon **AUG** & the promoter-operator region of this gene.

406. This sequence of 162 bases is called **leader sequence.**

407. The leader sequence at the **3' end of the gene** but 8-12 bases before the initiating codon, has a sequence **5' AGGAGGGA 3' (known as Shine Delgarno sequence).**

408. **Leader sequence of the phenylalanine operon** has three consecutive phenylalanine codon.

409. **Leucine operon**– 4 consecutive codon.

410. **Histidine operon**– 8 consecutive codon.

411. An important example of autogenous regulation is **p32** which is coded by **T_4 gene 32** and needed for recombination repair and replication of DNA.

412. Whenever this protein p32 is in excess, it binds to mRNA and prevents intiation of translation.

413. **Aptamers** are DNA or RNA molecules that have been selected from random pools based on their ability to bind other molecules including nucleic acid, protein and small organic compounds & even entire organism & have many potential uses in **medicine.**

414. The synthesis of transposase enzyme in **Tn10** is regulated through the synthesis of **anti-sense RNA.**

415. The regulation of gene expression in eukaryotes is exercised at the **post transcriptional level** involving RNA processing or translation. **(genome reorganization or gene rearrangement).**

416. The regulation of gene expression specific to eukaryotes is **chromosome remodelling.**

417. DNA sequences involved in regulation were also identified through the use of **DNA foot printing** as well as through the use of **gene constructs** for production of transgenic plant.

418. Silencer sequences have been reported for **human immune deficiency virus (HIV)** and called **negative response element (NRE)** and also for chicken lysosome gene.

419. A variety of specific DNA sequence (both upstream and downstream) involved in the regulation of gene expression has been identified. These sequences could be present either upstream or downstream and found associated with both the **house keeping genes or so called smart genes or luxury genes.**

These sequences include:

1. Promoter like TATA box (associated with CAAT box; GC box), BRE, Inr, DPE.
2. A variety of enhancers, silencers and insulators.

420. **Homeotic box in drosophila**

Homeotic box such that their mutation causes transformation of one body part into another.

421. These sequences (180 bps) found nearly 3' end of a gene in drosophila; name **Homeo box by Walter Gehring** of the University of **Basel in Switzerland.**

422. **DNA methylation** in eukaryotes mainly involves addition of a **methyl** group to the **carbon 5' position** of the cytosine ring.

423. The methylation is catalysed by **DNA methyl transferases** (*e.g.* DNMT1, DNMT 3a and **DNMT 3b in humans**) in the context of the sequences 5'-CG-3' which is also referred to as a CpG dinucleotide or CpG island.

424. Methylation leads to suppressing of gene expression while demethylation led to its activation.

425. Methylation sensitive enzymes – **Hpa II, MspII.**

426. Gene expression in certain tissues could also be induced through treatment with the **base analogue 5- azacytidine (5aza C)** which is known to inhibit methylation.

427. DNA methylation, thus, is the most common eukaryotic DNA modification which alters expression of genes and is therefore one of the many **epigenetic phenomenon** that are known to influence inheritance.

428. Repression of transcription from densely methylated genes can be mediated by **MeCP1**, and that from DNA sequence containing a single methylated CpG pair is mediated by **MeCP2.**

429. **Mechanism of transcription due to methylation-**

a) Blocking of transcription factor
b) Binding of specific transcriptional repressors
c) Chromatin modelling.

430. DNA methylation at **CpG dinucleotide** in mammals is essential for some important function such as **cell differentiation, imparting and X-inactivation.**

431. Several diseases are caused by defects within the methylation machinery, like the **Rett syndrome, fragile X syndrome and ICF (Immuno deficiency centromeric instability facial anomalies syndrome).**

432. CpG Island or CpG dinucleotide methylation is also capable of silencing tumor suppressor genes and may lead to increased mutation rate, which can occur frequently in these regions, leading to tumor development (cencer).

433. The study of transcription has been facilitated by a technique described as **NRD assay (Nuclear Run On assay** in which transcription continues in the presence of radio labled nucleotides) in the nuclei isolated from lysed cells.

434. The human β globin locus extends over 70 kb of DNA and contains an **LCR (Low control region)** & five **erythroid specific genes** ε-$^{G}\gamma^{A}$-γ-δ-β, arranged in the order of their developmental expression.

435. **Transvection** is a term that was coined in 1954 by **E. B. Lewis** for describing the phenomenon involving alteration in expression of a gene due to its pairing with its partner allele in the homologous chromosome **(E. B. Lewis died in 2004).**

436. Transvection in *Drosophila* is based on somatic chromosome pairing, **meiotic transvection** has also been reported in **neurospora.**

437. **The concept of post-transcriptional operon/regular** was proposed in 2002 by J. D. Keene of Duke University.

438. Antibody molecules **called immune globulins** are coded by genes, which undergo DNA rearrangements during cell differentiation, thus suggesting that although a genome usually remains constant in different cells, it does undergo a change with regards **to immune-globulin genes in mammals** (gene arrangement).

439. The term **"chromatin remodeling"** generally refers to change in histone-DNA interactions in nucleosomes.

440. Histone proteins are modified and non-histone chromosomal proteins.

441. Plants like Arabidopsis were also used for studies involving chromatin remodeling.

442. Component of chromatin remodeling complexes:-

a) **ATP dependent chromatin remodeling complex,** which make use of ATP for chromatin modification through mere histone DNA interactions.

b) **Histone modifying enzymes** which are used for post transcriptional modification of histones (mainly acetylation and methylation).

c) **Variants of the histone** (H2A, H2B, H3 & H4, H1) which are synthesized and incorporated into nucleosome sub-units.

443. ATP dependent **chromatin remodeling complexes** is also called **chromosome remodeling machines,** consists of ATP are polypeptide sub-unit and also non-ATPase sub-units.

Note:- **The genes that respond to interferon carry a common regulator gene or sequence called** ISRE (interferon stimulated response element).

444. **Lactoferrin** is an extracellular secretory protein, which in its dimeric or tetrameric form carries the message all by itself from cell surface to the cell nucleus.

445. The work on the role of **Nitric Oxide** as a signalling molecule was recognised in 1998 by the award of **Noble Prize** for medicine to the three scientists from USA (1) R.F. Frchgott, (2) Louis J. Hgnarro & (3) F. Murad.

446. Morphological and genetic markers have also been used for the study of development, particularly for the preparation of **fate maps** of early embryos.

447. The adult fly (Drosophila) consists of *(i)* **head** *(ii)* **three thoracic segment** (prothorax, mesothorax and metothorax denoted T1, T2, T3) *(iii)* **eight abdominals.**

448. AP2 (APETALA 2) gene is expressed in the leaf primodia, apical meristem and in floral primodia.

449. Floral meristem identify genes like **LFY, AP2, CAL actually work as bridge between floral induction genes and floral organ identity genes.**

450. **Restriction maps** can be used for determining the nucleotide sequence of a specific gene.

451. Mini-satellites and Micro-satellites/ SSRs or VNTRs has also been emphasized that poly allelic markers will be useful for mapping both the **simple mendelian traits as well as polygenic traits in segregating populations.**

452. Ribosomal-RNA genes were isolated for the 1st time by **H. Wallace and M. L. Birnstiel** in an amphibian named **Xenopus.**

453. Fredrick Sanger (who won noble prize twice) had initially developed a method for DNA sequencing, which utilized DNA polymerase to extend DNA chain length. This was termed **Plus-Minus Method.**

454. In a haploid set, the chromosomes are **non-homologous and have no-homologous to pair with,** they are found as univalent at metaphase 1 of meiosis.

455. In ***Oenothera Lamarckiana***, an American plant, giant mutation described by **Hugo de Vries** was later discovered to be an auto-tetraploid.

456. **Auto-triploids:** water melons, sugarbeet, tomato, grapes, banana etc.

457. **Auto-tetraploids:-** rye, corn, red clover, barseem, marigolds, snapdragons, phlox, grapes, apples etc.

458. Tetraploids X Diploid → Triploids.

459. Colchicine treatment used for developed triploids of sugarbeet, tomato & grapes.

Note:-

<u>DNA fingerprinting or forensic sciences:</u> The technique of DNA fingerprinting was developed for the 1st time by **Alec Jeffreys *et al.*** In this technique, DNA will be isolated from blood stains, semen stains or hair rooter & will be subjected to either **southern blotting & DNA hybridization with the help of specific probes or to PCR using PCR primers.**

460. **CDFD: Centre for DNA fingerprinting and diagnostics, Hyderabad.**

461. RBCs resemble as balushahi in shape.

462. In bacteria cells have range between 0.2 to 5.00 μm.

463. In ostrich, cell size of egg is 75 mm.

464. 6×10^{12} cells found in human beings.

465. Cell wall of a differentiated plant composed of three parts *e.g.*, middle lamella, primary cell wall and secondary cell wall.

466. Pectin is a major protein found in middle lamella.
467. Component of pectin in middle lamella in plant cell are calcium and magnesium salts.
468. Pectin is breakdown by pectinase or strong acids.
469. Cellulose is not found in middle lamella of cell wall.
470. Hemicelluloses are found between microfibrils of each layer and pectin is also present between two layer of microfibrils.
471. Primary cell wall contain maximum amount of hemicelluloses than other components
472. Secondary cell wall present between middle lamella and plasma lemma in plant.
473. Major component of secondary cell wall is cellulose.
474. Plasma lemma covered by cytoplasm of cell.
475. Plasma lemma consists of protein and lipids.
476. Glyco lipid present only in plasma lemma.
477. Endoplasmic reticulam is not present in egg, RBCs, undifferentiated cell and prokaryotes.
478. E. R. attatched with plasma lemma and nuclear envelope.
479. E. R. helps to organize base structure for protein synthesis.
480. Diameter of ribosome approx. 200 Å.
481. Ribosome composed of protein ans RNA.
482. The measurement of micro molecules by helps of S-unit (Sved berg unit).
483. Both animal and plant ribosome composed of 55 – 60% protein and 40 – 63% RNA.
484. In case of prokaryotes, ribosome consists of 37% protein and 63% RNA.
485. Mother of Golgi body is E. R.
486. Lysosome contain acid phosphatase, DNAse, acid RNAse, β-galactosidase.
487. Primary lysosome originates from Golgi complex.
488. Primary lysosome has hydrolytic enzyme.
489. Both food materials and hydrolytic enzymes found in secondary lysosome.
490. Lysosomes also called suicidal bags because its carry undigested portion of food materials.
491. Sherosome found only plant cell.
492. The discovery of sherosome was goes to Parner, 1953.
493. Newly synthesized vacuole (centriole) is also called pro centriole.
494. Centriole fond in animal cell.
495. Centrioles always present in pairs.
496. Kolliker, 1850 seem mitochondria in muscles.
497. Benda, 1897 proposed a term mitochondria.
498. Mitochondrion consists of protein, lipids, DNA and RNA.
499. The no. of oxysomes/mitochondria varies from $10^4 - 10^5$.

500. Oxisomes have ATPase enzyme and responsible for ATP synthesis.

501. Cellular furnances is also known as mitochondria or power house of cell or power plant of cell.

502. Enzymes responsible for synthesis of fatty acids found in inner chamber or matrix of mitochondria.

503. The term plastid was proposed by E. Haeckel, 1866.

504. Amyloplast is also called Leucoplast.

505. Leucoplast stored starch, lipid and protein.

506. Chromoplast does not have chlrorophyll.

507. Chromoplast is also helps to synthesize of membrane lipids.

508. Chromoplast color is radish or yellow due to presence of carotenoid pigments.

509. Chroloplast composed of protein, lipid, chlorophyll, carotenoids, RNA and DNA.

510. Chlorophyll and carotenoid molecules are both associated with thylakoid membrane of chloroplast.

511. Range of grana in one chloroplast is 40 to 60.

512. Quantasomes are the unit of photosynthetic present on the surface of grana.

513. Quantasome consists of 250 to 300 molecules of chlorophyll.

514. Park and Biggins discovered Quantasomes.

515. Glyoxisomes are a membranous structure contain various enzymes for β-oxidation of fatty acid and glyoxilate path way.

516. The enzymes are responsible for β-oxidation of fatty acid and glyoxilate path way found in glyoxisomes were discovered by Briedenbach, 1967.

517. The diameter of glyoxisomes is varies from 0.5 to 1.5 μm.

518. The β-oxidation of fatty produces Acetyl Co A during glyoxylate cycle.

519. The term ribosome was invented by Palade, 1958.

520. Highly folded circular DNA molecules do not have histone protein as referred as nucleoid.

521. Sacroplasmic reticulum stores calcium.

522. Ribosome may not be present in erythrocytes.

523. Spherosome do not carry phospholipids bilayer structure.

524. 70's ribosomes are present in cytosol.

525. The arrangement of fibrils in centrioles is 9 + 0.

526. Chromoplast is helps in pollination, seed dispersal, and synthesis of membrane lipids.

527. E. R. anxious with glycosylation of protein.

528. Inner membrane of mitochondria has oxysomes.

529. Golgi complex comes up from endoplasmic reticulum.

530. Blue green algae do not have golgi complex.

531. Conversion of pyruvate and fatty acid into Acetyl Co A during Kreb's cycle which have major source of energy is NADH.

532. A true fact, a chromosome has many genes.

533. Independently assortment represent the independently segregation of different forms of alleles or genes present in pairs of homologous chromosomes during meiosis.

534. Segregation refers the separation of genes or alleles from each other.

535. Recombination phenotypes consist of all new combination of inherited visual characters.

536. Crossing over refers that the gametes contain recombinant combination of various genes or markers located same chromosomes are produced.

537. Genes do not exhibited independently assort during crossing over; such genes called linked genes.

538. Transmission of genes that they do not show independent assortment, called linkage.

539. The effects of linkage were Ist reported by W. Bateson and Punnett in 1906 working on sweet pea.

540. Coupling or cis configuration of genes refers that the arrangement of mutant and wild types genes together in heterozygous form.

541. Wild alleles are also show normal or dominant form than recessive.

542. Position on a chromosomes that carry genes, called locus.

543. Arrangement of loci on chromosomes is linear sequence.

544. Breakage and exchange of chromosomal segment involves in crossing over process.

545. Crossing over occurs after synapsis during pachytene phase of prophase.

546. Interphase carries meiotic crossing over.

547. Post replication tetrad stature of chromosome comes after meiotic crossing over.

548. Chiasma frequencies were positive correlated with recombinant frequencies.

549. Direct relationship occurs between cross over and chiasmata.

550. Study on the cytological basis that homologous chromosomes exchange part during crossing over was first reported by C. Stern in 1931 working on Drosophilla.

551. In plant kingdom, cyto-basis of crossing over was 1st reported by H. B. Creighton and McClintock in working on maize.

552. H. B. Creighton and McClintock were study of chromosomes no 9 in maize to given cytological basis of crossing over.

553. Recombination refers that the formation of new combination.

554. Drossophilla has four pairs of linkage group.

555. Maize has ten pairs of linkage group.

556. Mouse has twenty pairs of linkage group.

557. Circular linkage map seen in viruses, prokaryotes and mitochondria of eukaryotic cells.

558. Linkage map frequencies were observed by the help of the frequencies of cross over or recombinant chromosomes.

559. Each meiotic cross over event yields two cross over or recombination.

560. Ranges of the frequencies of recombinant are 0 to 50%.

561. Ratio between two strand to three strand to four strand crossing over will be obtained as 1 : 2 : 1. Deviation from this ratio is called to the results of chromatid interferences.

562. Models of crossing over theory were based on the assumption of conservative DNA synthesis.

563. A small amount of copy choise DNA repair synthesisis associated with crossing over through breakage and reunion.

564. Breakage and reunion system of DNA is responsible for occurrence of nonparental recombination of gene conservation.

565. Synaptinemal complexes have not been observed in Drosophilla males.

566. Less than 1% DNA synthesis occurs at the time of synaptinemal complex formation.

567. Popular model of crossing over based on Holliday model and the Whitehouse model.

568. Mechanism of crossing over is also based on breakage and reunion that associated with DNA repairs.

569. Endo-nuclease breaks the single strand of each of the two parental DNA molecules.

570. Helix destabilization protein and DNA unwinding protein is also known as DNA helicase.

571. Rec A protein is responsible to increment of renaturation rate by over 50 folds.

572. DNA ligase is a enzyme responsible to rejoining of two strand of DNA.

573. A double strand break model of crossing over was 1st proposed by Szostak *et al.*, in 1983.

574. Exonuclease enzyme is cut the DNA at terminal portion.

575. Illegitimate recombination refers to the recombination of non-homologous DNA molecule.

576. Nucleoid of a bacterium contains two or more identical copies of the chromosomes and plasmids.

577. Transformation was 1st proposed by Griffiths in 1928 working on pathogenic strains of *Deplococcus pneumonia.*

578. The work of Griffith (1928) was proofed by three scientists Avery, Macleod and McCarty in 1944.

579. Transduction was 1st discovered by N. Zinder and J. Lederberg in 1952.

580. Transduction by which a bacteriophase particle carries a segment of the chromosome from one bacterium to other.

581. During transduction, transducing segment of chromosomes particles are produced very low frequency.

582. trp gene are require for the synthesis of the amino acid trypophan in *E.coli.*

583. Formation of specialized transducing segment or particles during pro-phage of excision.

584. Conjugation was discovered by Leaderberg J. and Tatum E. L. in 1946.
585. Both Leaderberg and Tatum were received Noble Prize in 1958.
586. A donor cell containing the F-factor, cell known as F^+ cell.
587. F-Factor has an autonomous property.
588. A unit of genetic material has a special capacity to replication of independently, called plasmid.
589. Plasmid is also a replicon.
590. Plasmids are dispensible.
591. Plasmids are found in bacteria.
592. Plasmids encode a toxin substance called bacteriocins or colicins.
593. Vibriocins present in bacteria has capable to kill the sensitive *Vibrio cholera* cells.
594. The recombinant joint has the ability to move along the DNA duplexes. This type movement known as branch migration.
595. Branch migration may be catalyzed by a recombinant enzyme.
596. DNA duplexes are joined by DNA ligase.
597. A strand of DNA duplex in which break is introduced, called recipient duplex.
598. Double strand break in yeast are formed by Spo 11.
599. A event like branch migration, DNA synthesis and resolution of recombinant joints occurs during phechytene.
600. Yeast mutation, chromosomes pairing allows by Zip 2 protein.
601. In yeast, zip2 protein prevents the formation of synaptinemal complex.
602. In yeast, the specificity of chromosomes pairing controlled by Hop 2 proteins.
603. Hop 2 protein leads the pairing of non homologous chromosomes.
604. Ruv B is a type of ATPase lead helicase activity and provides branch migration.
605. Rec G protein lead similar activity of Ruv B.
606. Gene conversion is correlated with crossing over.
607. The term "Genetics" was 1st invented by Bateson in 1905.
608. The term "Genetics" was derived from Greak word "gene" meaning to be born.
609. Gene was a smallest unit that could undergo a mutational change or could take a partin recombination.
610. Gene is a unit of mataion does not carry information generation to generation.
611. Sturtevant (1925) published his work on analysis of Bar eye mutant in *Drosophilla.*
612. In Drosophilla, a recessive sex linked mutant called lozenge (*Lz*).
613. *Lz* gene is lead the formation of smaller, darken and more elliptically shaped eyes.
614. The father of fine structure of gene and pseudo-allelism is E. B. Levis.
615. E. B. Levis (1952) provided a classical example in form of a cross of apricot eye x white eye produce F1 flies having intermediate color.
616. Benzer was classified all types of mutant in two catagories A-revertant tupe B-non revertant types.
617. Benzer observed deletion in segments of different size inn rII region.
618. Non mendelian inheritance is also known as extra-chromosomal inheritance.
619. Non mendelian inheritance involves DNA replication in cytoplasmic organelles *e.g.*, mitochandria and plastid

620. Symbiont bacteria were discovered in the cytoplasm of the protozoan (*Paramecium aurelia*).
621. French W. L. was observed that sterility in hybrid *Culex mosquitoes* caused by mt DNA.
622. Human, mice, cattle each have 16569 mt DNA sequenced reported 1st time.
623. Corren's observed significantly results of reciprocal crosses in 1908.
624. Corren's was 1st proposed non Mendelian inheritance in Four 'O' clock plant.
625. Scientific name of Four 'O' clock plant is *Mirabilis jalapa.*
626. Proplastid is a cypoplasmic particles produced chloroplasts.
627. Chloroplasts contain cpDNA.
628. Chloroplast has capacity to replicating themselves.
629. Majority of mt Protein must be encoded by nuclear genes.
630. CpDNA gene number are present in tobacco is 150.
631. A ways of self fertilization called autogamy.
632. Autogamy leads homozygosis in population.
633. Beale G. H. was discovered erythromycin resistance in *Paramecium* caused by non Mendelian inheritance.
634. Extra chromosomal DNA molecules have a property of self replicating.
635. A plasmid called Ti that carries a DNA sequence that transforms cells of dicotyledonous plant to tumor cells.
636. Tumor formation is associated with crown gall disease.
637. Crown gall disease caused by *Agrobacterium tumefancies.*
638. Maternal effects are controlled by nuclear genes of the mother parents.
639. Out most layer of endosperm is also known as aleurone.
640. Grande is defines as a petite character is artificial induced in wild type and carrying several mitochondria markers genes.
641. Nucleic acid was 1st called nuclein.
642. Term nuclein is a term for nucleic acid because its isolates from cell nuclei.
643. The term nuclein was 1st proposed by Miescher F.
644. Micro-molecules composed a repeating form of nucleic acid called nucleotides.
645. Purines are as adenine and guanine.
646. Purines have double ring stature.
647. Pyrimidines are as thymine, uracil and cytosine.
648. Thymine, uracil and cutosine carry single ring stature.
649. True stature of DNA was 1st deduced by Watson J. D. and Crick F. H. C. in 1953.
650. In DNA, adenine is always bound to thymine.
651. In DNA, guanine is always bound to cytosine.
652. In double helix DNA, adenine is always equal to thymine.
653. In double helix DNA, cytosine is always equal to guanine.
654. Double helix form of DNA was proposed by Watson and Crick.
655. In case of high salts concentration, DNA seems as in A form.
656. A form of DNA caries eleven nucleotide pairs per turns.
657. One Å is equal to 10^{-8}.

658. A polynucleotide chain consists of many nucleotide liked by phosphodiester bond.
659. The two polynucleotide strands are held each other in their helical configuration by helping of hydrogen bond.
660. The two strands of a DNA double helix are showed opposite form (non identical).
661. In DNA double helix has their backbone made by the joining of sugar and phosphate group.
662. The complementary polarity of the DNA strands is very useful in regarding the mechanism of replication of DNA.
663. The sugar attached to be base, results nucleoside.
664. Nucleoside may be changed to nucleotide by attachment of phosphoric acid to the 5′ carbon of the sugar.
665. In DNA, the phosphate group is denoted by β, α, and γ.
666. In DNA, α group of phosphate is attached direct to sugar.
667. A simplest DNA packaging stature present in all eukaryotic chromosomes called nucleosomes.
668. DNA is wrapped around an octamer of small basic protein called histones protein.
669. H2A, H2B, H3 and H4 are known as core-histones protein.
670. DNA helix coiled in nucleus.
671. J. Cairn's (1963) was 1st invented the technique of auto-radiography for revelation of replicating chromosome.
672. Radioautography is a method for detecting & localizing radioactive isotopes in cytological preparations or macromolecules by exposure to a photographic emulsion that is sensitive to **low energy radiation.**
673. Autoradiograhy is particularly used in studying DNA metabolism because **DNA can be specially labelled on growing cells with thymidine.**
674. Cairns interpretation of the autograph was that semiconservative replication started at a site on the chromosome, which he called **origin** & proceeded sequentially & unidirectionally around the circular structure.
675. Each Y shaped structure is a replication fork and the two replication forks move opposite directions sequentially around the circular chromosome.
676. The A-T rich clusters were used as physical markers by M. Schnos and R.B. Mman to demonstrate, using a technique called **denaturation mapping.**
677. When DNA molecules are exposed to high temperature (100%) or high pH (11.4), then hydrogen & hydrophobic bonds that hold the complementary strand together in a double-helix configuration get broken down resulting in separation of two strands. This process is called **denaturation.**
678. A = T rich molecule is easily denaturate than G ≡ C.
679. When Lambda chromosomes are exposed to pH 10.05 for 10 minutes under appropriate conditions, the A-T rich clusters sets denatured and forms "denaturation bubbles" which are detectable by electron microscopy and can be used as physical markers irrespective of Lambda chromosome.
680. Replication of DNA molecules in the chromosome of eukaryotes is also **bidirectional.**

❍❍❍